ENGINEERING FUNDAMENTALS
QUICK REFERENCE CARDS

Fourth Edition

Michael R. Lindeburg, P.E.

PROFESSIONAL PUBLICATIONS, INC.
Belmont, CA 94002

In the ENGINEERING REFERENCE MANUAL SERIES

Engineer-In-Training Reference Manual
  Engineering Fundamentals Quick Reference Cards
  Engineer-In-Training Sample Examinations
  Mini-Exams for the E-I-T Exam
  1001 Solved Engineering Fundamentals Problems
  E-I-T Review: A Study Guide
Civil Engineering Reference Manual
  Civil Engineering Quick Reference Cards
  Civil Engineering Sample Examination
  Civil Engineering Review Course on Cassettes
  Seismic Design of Building Structures
  Seismic Design Fast
  Timber Design for the Civil P.E. Exam
246 Solved Structural Engineering Problems
Mechanical Engineering Reference Manual
  Mechanical Engineering Quick Reference Cards
  Mechanical Engineering Sample Examination
  101 Solved Mechanical Engineering Problems
  Mechanical Engineering Review Course on Cassettes
  Consolidated Gas Dynamics Tables
Electrical Engineering Reference Manual
  Electrical Engineering Quick Reference Cards
  Electrical Engineering Sample Examination
Chemical Engineering Reference Manual
  Chemical Engineering Quick Reference Cards
  Chemical Engineering Practice Exam Set
Land Surveyor Reference Manual
Petroleum Engineering Practice Problem Manual
Expanded Interest Tables
Engineering Law, Design Liability, and Professional Ethics
Engineering Unit Conversions

In the ENGINEERING CAREER ADVANCEMENT SERIES

How to Become a Professional Engineer
The Expert Witness Handbook—A Guide for Engineers
Getting Started as a Consulting Engineer
Intellectual Property Protection—A Guide for Engineers
E-I-T/P.E. Course Coordinator's Handbook
Becoming a Professional Engineer

ENGINEERING FUNDAMENTALS QUICK REFERENCE CARDS
Fourth Edition

Printed in the United States of America

ISBN: 0-912045-34-5

Professional Publications, Inc.
1250 Fifth Avenue, Belmont, CA 94002
(415) 593-9119

Current printing of this edition (last number): 6 5 4 3 2 1

TABLE OF CONTENTS

FOR INSTANT RECALL 1
CONVERSION FACTORS 2
FUNDAMENTAL AND PHYSICAL CONSTANTS 3
MATHEMATICS 4
ENGINEERING ECONOMICS 8
FLUID STATICS AND DYNAMICS 10
THERMODYNAMICS 13
POWER CYCLES AND REFRIGERATION 16
CHEMISTRY 18
STATICS 20
MATERIALS SCIENCE AND TESTING 22
MECHANICS OF MATERIALS 25
DYNAMICS 28
DC ELECTRICITY 30
AC ELECTRICITY 32
PERIPHERAL SCIENCES 34
MODELING OF ENGINEERING SYSTEMS 35
ATOMIC AND NUCLEAR THEORY 36
INDEX 38

FOR INSTANT RECALL

FUNDAMENTAL CONSTANTS

N_A Avogadro's number: 6.022×10^{23} mol^{-1}
R^* universal gas constant: 1545 ft-lbf/lbmol-°R; 8314 J/kmol·K
c speed of light: 9.84×10^8 ft/sec; 3×10^8 m/s
g_c gravitational constant: 32.174 lbm-ft/lbf-sec^2
J Joule's constant: 778.17 ft-lbf/BTU

TEMPERATURE CONVERSIONS (21-4)

(Use of the degree symbol with the Kelvin temperature scale is not standard.)

$$°F = 32 + \left(\tfrac{9}{5}\right) °C$$
$$°C = \left(\tfrac{5}{9}\right) (°F - 32)$$
$$°R = °F + 460$$
$$°K = °C + 273$$
$$\Delta°R = \Delta°F = \left(\tfrac{9}{5}\right) \Delta°K = \left(\tfrac{9}{5}\right) \Delta°C$$
$$\Delta°K = \Delta°C = \left(\tfrac{5}{9}\right) \Delta°R = \left(\tfrac{5}{9}\right) \Delta°F$$

DERIVED DATA AND PHYSICAL PROPERTIES

atmospheric pressure: 14.696 psia; 29.92 in of mercury; 407.1 in of water; 1.0133×10^5 Pa
OR 101.33 kPa

circle: 360°; 2π rad

density of air at 1 atm and 70°F: 0.075 lbm/ft^3; 1.20 kg/m^3

density of mercury: 0.491 lbm/in^3; 1.360×10^4 kg/m^3

density of water: 62.4 lbm/ft^3; 0.0361 lbm/in^3; 1 g/cm^3; 1000 kg/m^3

specific gravity:

mercury	13.6
water	1.0

frequency of house current: 60 Hz; 377 rad/s

standard gravity: 32.17 ft/sec^2; 386 in/sec^2; 9.807 m/s^2

modulus of elasticity for steel: 2.9×10^7 psi; 2×10^{11} Pa

modulus of shear for steel: 1.15×10^7 psi; 7.9×10^{10} Pa

molecular weight:

air	29.0
carbon	12.0
carbon dioxide	44.0
helium	4.0
hydrogen	2.0
nitrogen	28.0
oxygen	32.0

ratio of specific heats for air: 1.4

specific gas constant for air: 53.3 ft-lbf/lbm-°R; 287 J/kg·K

approximate specific heats:

ice	0.5 BTU/lbm-°F; 2 kJ/kg·K
water	1.0 BTU/lbm-°F; 4.2 kJ/kg·K
steam	0.5 BTU/lbm-°F; 2 kJ/kg·K

triple point of water: 32.02°F, 0.0888 psia; 273.34K, 0.00592 atm

FORMULAS

area of a circle $\quad A = \pi r^2 = \frac{\pi}{4} d^2$

circumference of a circle $\quad p = 2\pi r = \pi d$

area of a triangle $\quad A = \frac{1}{2} bh$

volume of a sphere $\quad V = \frac{4}{3}\pi r^3 = \frac{\pi}{6} d^3$

area moment of inertia for a rectangle

$$I_{\text{centroid}} = \frac{1}{12} bh^3$$
$$I_{\text{side}} = \frac{1}{3} bh^3$$

area moment of inertia for a circle

$$I_{\text{centroid}} = \frac{1}{4}\pi r^4 = \frac{\pi}{64} d^4$$

polar moment of inertia for a circle

$$J = \frac{1}{2}\pi r^4 = \frac{\pi}{32} d^4$$

mass moment of inertia of cylinder

$$J = \frac{1}{2} m r^2 = \frac{1}{8} m d^2$$

pressure and head

$$p = \gamma h = \rho h \times \frac{g}{g_c}$$

rotational speed

$$\omega = 2\pi f = 2\pi \left(\frac{\text{rpm}}{60}\right)$$

decibels

$$\text{dB} = 10 \log_{10} \left(\frac{P_2}{P_1}\right)$$

CONVERSION FACTORS (A-1)

to convert	into	multiply by
acres	square feet	43,560.0
acres	square miles	1.562×10^{-3}
ampere-hours	coulombs	3600.0
angstrom units	inches	3.937×10^{-9}
angstrom units	microns	1×10^{-4}
astronomical units	kilometers	1.495×10^{8}
atmospheres	centimeters of mercury	76.0
atmospheres	inches of mercury	29.92
BTU	foot-pounds	778
BTU	horsepower-hours	3.931×10^{-4}
BTU	kilowatt-hours	2.928×10^{-4}
BTU/hour	watts	0.2931
calories	BTU	3.9685×10^{-3}
centimeters	kilometers	1×10^{-5}
centimeters	meters	1×10^{-2}
centimeters	millimeters	10.0
centimeters	feet	3.281×10^{-2}
centimeters	inches	0.3937
coulombs	faradays	1.036×10^{-5}
cubic centimeters	cubic inches	0.06102
cubic centimeters	pints (U.S. liquid)	2.113×10^{-3}
cubic feet	cubic meters	0.02832
cubic feet	gallons	7.48
cubic feet/minute	pounds water/minute	62.43
cubic feet/second	gallons/minute	448.83
days	seconds	86,400.0
degrees (angle)	radians	1.745×10^{-2}
degrees/second	revolutions/minute	0.1667
dynes	newtons	1×10^{-5}
ergs	BTU	9.480×10^{-11}
ergs	foot-pounds	7.3670×10^{-8}
ergs	kilowatt-hours	2.778×10^{-14}
faradays/second	amperes	96,500
feet	centimeters	30.48
feet	meters	0.3048
feet	miles (nautical)	1.645×10^{-4}
feet	miles (statute)	1.894×10^{-4}
feet/minute	centimeters/second	0.5080
feet/second	knots	0.5921
feet/second	miles/hour	0.6818
foot-pounds	BTU	1.286×10^{-3}
foot-pounds	kilowatt-hours	3.766×10^{-7}
gallons	cubic feet	0.1337
gallons	liters	3.785
gallons of water	pounds of water	8.3453
gallons/minute	cubic feet/hour	8.0208
gallons/minute	cubic feet/second	0.00223
grams	ounces (avoirdupois)	3.527×10^{-2}
grams	ounces (troy)	3.215×10^{-2}
grams	pounds	2.205×10^{-3}
horsepower	BTU/minute	42.44
horsepower	foot-pounds/minute	33,000
horsepower	foot-pounds/second	550
horsepower	kilowatts	0.7457
horsepower	watts	745.7
hours	days	4.167×10^{-2}
hours	weeks	5.952×10^{-3}
inches	centimeters	2.540
inches	miles	1.578×10^{-5}
joules	BTU	9.480×10^{-4}
joules	ergs	1×10^{7}
kilograms	pounds	2.205
kilograms	slugs	0.068522
kilometers	feet	3281.0
kilometers	meters	1000.0
kilometers	miles	0.6214
kilometers/hour	knots	0.5396

to convert	into	multiply by
kilopascals	pounds/square inches	0.145
kilowatts	foot-pounds/second	737.6
kilowatts	horsepower	1.341
kilowatt-hours	BTU	3413.0
knots	feet/hour	6080.0
knots	nautical miles/hour	1.0
knots	statute miles/hour	1.151
light years	miles	5.9×10^{12}
links (surveyor's)	inches	7.92
liters	cubic centimeters	1000.0
liters	cubic inches	61.02
liters	gallons (U.S. liquid)	0.2642
liters	milliliters	1000.0
liters	pints (U.S. liquid)	2.113
meters	centimeters	100.0
meters	feet	3.281
meters	kilometers	1×10^{-3}
meters	miles (nautical)	5.396×10^{-4}
meters	miles (statute)	6.214×10^{-4}
meters	millimeters	1000.0
microns	meters	1×10^{-6}
miles (nautical)	feet	6080.27
miles (statute)	feet	5280.0
miles (nautical)	kilometers	1.853
miles (statute)	kilometers	1.609
miles (nautical)	miles (statute)	1.1516
miles (statute)	miles (nautical)	0.8684
miles/hour	feet/minute	88.0
miles/hour	feet/second	1.467
milligram/liter	parts/million	1.0
milliliters	liters	1×10^{-3}
millimeters	inches	3.937×10^{-2}
newtons	dynes	1×10^{5}
newtons	pounds	0.2248
ounces	pounds	6.25×10^{-2}
ounces (troy)	ounces (avoirdupois)	1.09714
parsecs	miles	1.92×10^{13}
parsecs	kilometers	3.084×10^{13}
pints (liquid)	cubic centimeters	473.2
pints (liquid)	cubic inches	28.87
pints (liquid)	gallons	0.125
pints (liquid)	quarts (liquid)	0.5
pounds	kilograms	0.4536
pounds	ounces	16.0
pounds	ounces (troy)	14.5833
pounds	pounds (troy)	1.21528
quarts (dry)	cubic inches	67.20
quarts (liquid)	cubic inches	57.75
quarts (liquid)	gallons	0.25
quarts (liquid)	liters	0.9463
radians	degrees	57.30
radians	minutes	3438.0
revolutions	degrees	360.0
revolutions/minute	degrees/second	6.0
seconds	minutes	1.667×10^{-2}
slugs	pounds	32.17
tons (long)	kilograms	1016.0
tons (short)	kilograms	907.18
tons (long)	pounds	2240.0
tons (short)	pounds	2000.0
tons (long)	tons (short)	1.120
tons (short)	tons (long)	0.89287
watts	BTU/hour	3.4129
watts	horsepower	1.341×10^{-3}
yards	meters	0.9144
yards	miles (nautical)	4.934×10^{-4}
yards	miles (statute)	5.682×10^{-4}

quantity	symbol	English	SI
Charge			
electron	e		-1.6022×10^{-19} C
proton	p		$+1.6021 \times 10^{-19}$ C
Density			
air [STP]		0.0805 lbm/ft^3	1.29 kg/m^3
air [70°F (20°C), 1 atm]		0.0749 lbm/ft^3	1.20 kg/m^3
earth [mean]		345 lbm/ft^3	5520 kg/m^3
mercury		849 lbm/ft^3	1.360×10^4 kg/m^3
sea water		64.0 lbm/ft^3	1025 kg/m^3
water [mean]		62.4 lbm/ft^3	1000 kg/m^3
Distance [mean]			
earth radius		2.09×10^7 ft	6.370×10^6 m
earth-moon separation		1.26×10^9 ft	3.84×10^8 m
earth-sun separation		4.89×10^{11} ft	1.49×10^{11} m
moon radius		5.71×10^6 ft	1.74×10^6 m
sun radius		2.28×10^9 ft	6.96×10^8 m
first Bohr radius	a_0	1.736×10^{-10} ft	5.292×10^{-11} m
Gravitational Acceleration			
earth [mean]	g	32.174 (32.2) ft/sec^2	9.8067 (9.81) m/s^2
moon [mean]		5.47 ft/sec^2	1.67 m/s^2
Mass			
atomic mass unit	u	3.66×10^{-27} lbm	1.6606×10^{-27} kg
earth		4.11×10^{23} slugs	6.00×10^{24} kg
earth [customary U.S.]		1.32×10^{25} lbm	n.a.
electron [rest]	m_e	2.008×10^{-30} lbm	9.109×10^{-31} kg
moon		1.623×10^{23} lbm	7.36×10^{22} kg
neutron [rest]	m_n	3.693×10^{-27} lbm	1.675×10^{-27} kg
proton [rest]	m_p	3.688×10^{-27} lbm	1.673×10^{-27} kg
sun		4.387×10^{30} lbm	1.99×10^{30} kg
Pressure, atmospheric		14.696 (14.7) lbf/in^2	1.0133×10^5 Pa
Temperature, standard		32°F (492°R)	0°C (273K)
Velocity			
earth escape		3.67×10^4 ft/sec	1.12×10^4 m/s
light [vacuum]	c	9.84×10^8 ft/sec	2.9979 (3.00) $\times 10^8$ m/s
sound [air, STP]	a	1090 ft/sec	331 m/s
[air, 70°F (20°C)]		1130 ft/sec	344 m/s
Volume, molal ideal gas [STP]		359 ft^3/lbmol	22.41 m^3/kmol
Fundamental Constants			
Avogadro's number	N_A		6.022×10^{23} mol^{-1}
Bohr magneton	μ_B		9.2732×10^{-24} J/T
Boltzmann constant	κ	5.65×10^{-24} ft-lbf/°R	1.3807×10^{-23} J/K
Faraday constant	F		96 487 C/mol
gravitational constant	g_c	32.174 lbm-ft/lbf-sec^2	
gravitational constant	G	3.44×10^{-8} ft^4/lbf-sec^4	6.672×10^{-11} N·m^2/kg^2
nuclear magneton	μ_N		5.050×10^{-27} J/T
permeability of a vacuum	μ_0		1.2566×10^{-6} N/A^2 (H/m)
permittivity of a vacuum	ϵ_0		8.854×10^{-12} C^2/N·m^2 (F/m)
Planck's constant	h		6.6256×10^{-34} J·s
Rydberg constant	R_∞		1.097×10^7 m^{-1}
specific gas constant, air	R	53.3 ft-lbf/lbm-°R	287 J/kg·K
Stefan-Boltzmann constant		1.71×10^{-9} BTU/ft^2-hr-°R^4	5.670×10^{-8} W/m^2·K^4
triple point, water		32.02°F, 0.0888 psia	0.01109°C, 0.6123 kPa
universal gas constant	R^*	1545 ft-lbf/lbmol-°R	8314 J/kmol·K
	R^*	1.986 BTU/lbmol-°R	

MATHEMATICS

QUADRATIC EQUATION (3-3)

The roots of the equation $ax^2 + bx + c = 0$ are

$$\frac{-b \pm \sqrt{b^2 - 4ac}}{2a}$$

POLYNOMIALS (3-3)

$$(a+b)(a-b) = a^2 - b^2$$

$$(a \pm b)^2 = a^2 \pm 2ab + b^2$$

$$(a \pm b)^3 = a^3 \pm 3a^2b + 3ab^2 \pm b^3$$

$$(a^3 \pm b^3) = (a \pm b)(a^2 \mp ab + b^2)$$

$$(a^n - b^n) = (a-b)(a^{n-1} + a^{n-2}b + a^{n-3}b^2 + \cdots + b^{n-1}) \quad [n \text{ is any positive integer}]$$

$$(a^n + b^n) = (a+b)(a^{n-1} - a^{n-2}b + a^{n-3}b^2 - \cdots + b^{n-1}) \quad [n \text{ is any positive odd integer}]$$

BINOMIAL SERIES (3-3)

$$(a+b)^n = a^n + na^{n-1}b + C_2a^{n-2}b^2 + \cdots + C_{i-1}a^{n+1-i}b^{i-1} + \cdots + nab^{n-1} + b^n$$

$$C_i = \frac{n!}{i!(n-i)!} \qquad i = 0, 1, 2, \ldots n$$

EXPONENTS (3-4)

$$b^0 = 1 \quad [b \neq 0]$$

$$b^1 = b$$

$$b^{-n} = \frac{1}{b^n} = \left(\frac{1}{b}\right)^n \quad [b \neq 0]$$

$$\left(\frac{a}{b}\right)^n = \frac{a^n}{b^n} \quad [b \neq 0]$$

$$(ab)^n = a^nb^n$$

$$b^{m/n} = \sqrt[n]{b^m} = \left(\sqrt[n]{b}\right)^m$$

$$(b^n)^m = b^{nm}$$

$$b^mb^n = b^{m+n}$$

$$\frac{b^m}{b^n} = b^{m-n} \quad [b \neq 0]$$

$$\sqrt[n]{b} = b^{1/n}$$

$$\left(\sqrt[n]{b}\right)^n = \left(b^{1/n}\right)^n = b$$

$$\sqrt[n]{ab} = \sqrt[n]{a}\,\sqrt[n]{b} = a^{1/n}b^{1/n} = (ab)^{1/n}$$

$$\sqrt[n]{\frac{a}{b}} = \frac{\sqrt[n]{a}}{\sqrt[n]{b}} = \left(\frac{a}{b}\right)^{1/n} \quad [b \neq 0]$$

$$\sqrt[m]{\sqrt[n]{b}} = \sqrt[mn]{b} = b^{1/mn}$$

LOGARITHM IDENTITIES (3-5)

$$x^a = \text{antilog}[a \log(x)]$$

$$\log(x^a) = a \log(x)$$

$$\log(xy) = \log(x) + \log(y)$$

$$\log\left(\frac{x}{y}\right) = \log(x) - \log(y)$$

$$\ln(x) = \frac{\log_{10}x}{\log_{10}e} \approx 2.3(\log_{10}x)$$

$$\log_b(b) = 1$$

$$\log(1) = 0$$

$$\log_b(b^n) = n$$

$$\ln e^x = x$$

COMPLEX NUMBERS (3-7) (ϕ in radians)

$$a + bi \equiv r\angle\theta \equiv r(\cos\theta + i\sin\phi) \equiv re^{i\theta}$$

$$a - bi \equiv r\angle\theta \equiv r(\cos\theta - i\sin\phi) \equiv re^{-i\theta}$$

$$r = \sqrt{a^2 + b^2}$$

$$\theta = \arctan\left(\frac{b}{a}\right)$$

EULER'S EQUATION (3-7)

$$e^{i\theta} = \cos\theta + i\sin\theta$$

$$e^{-i\theta} = \cos\theta - i\sin\theta$$

$$\cos\theta = \frac{e^{i\theta} + e^{-i\theta}}{2}$$

$$\sin\theta = \frac{e^{i\theta} - e^{-i\theta}}{2i}$$

GEOMETRIC SERIES (3-11)

A geometric series is given by

$$a + ar + ar^2 + \cdots + ar^{n-1}$$

If n is finite, the series converges. The nth term and sum are

$$a_n = ar^{n-1}$$

$$S_n = \frac{a(1 - r^n)}{1 - r}$$

If n is infinite, the series converges only for $-1 < r < 1$. In this case, the sum of an infinite number of terms is

$$S_n = \frac{a}{1 - r}$$

The series diverges if $|r| \geq 1$.

ARITHMETIC SERIES (3-11)

An arithmetic series is given by

$$a + (a + d) + (a + 2d) + \cdots$$

If the series has a finite number of terms, it converges. If the series has an infinite number of terms, it diverges. The nth term and sum of n terms are

$$a_n = a + (n-1)d$$

$$S_n = \tfrac{1}{2}n[2a + (n-1)d]$$

DETERMINANT (4-2)

For a second-order determinant,

$$\begin{vmatrix} a_1 & a_2 \\ b_1 & b_2 \end{vmatrix} = a_1b_2 - a_2b_1$$

For a third-order determinant,

$$\begin{vmatrix} a_1 & a_2 & a_3 \\ b_1 & b_2 & b_3 \\ c_1 & c_2 & c_3 \end{vmatrix} = a_1b_2c_3 + a_2b_3c_1 + a_3b_1c_2 - c_1b_2a_3 - c_2b_3a_1 - c_3b_1a_2$$

VECTORS (Chap. 5)

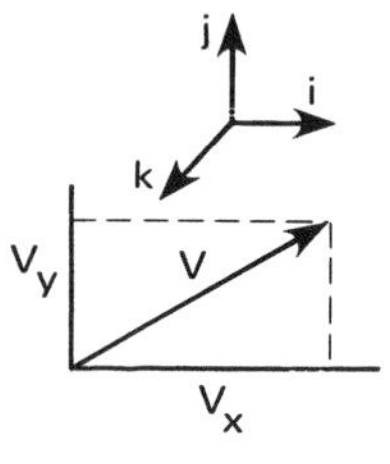

A vector is a directed line segment. It is defined completely only when both the magnitude and direction are known. Vectors are defined in terms of the unit vectors **i**, **j**, **k**.

The unit vectors are vectors of unit length directed along the x, y, and z axes, respectively. A vector, **V**, emanating from the origin, can be written in terms of the unit vectors and its endpoint (V_x, V_y, V_z).

$$\mathbf{V} = V_x\mathbf{i} + V_y\mathbf{j} + V_z\mathbf{k}$$

Addition of vectors is performed by adding the components

$$\mathbf{A} + \mathbf{B} = (A_x + B_x)\mathbf{i} + (A_y + B_y)\mathbf{j} + (A_z + B_z)\mathbf{k}$$

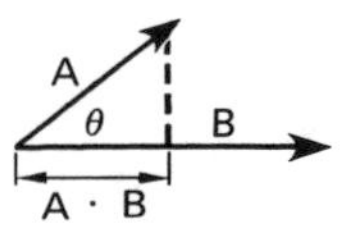

The *dot product* is a scalar and is the length of the projection of **B** onto **A**.

$$\mathbf{A}\cdot\mathbf{B} = A_xB_x + A_yB_y = |\mathbf{A}||\mathbf{B}|\cos\theta$$

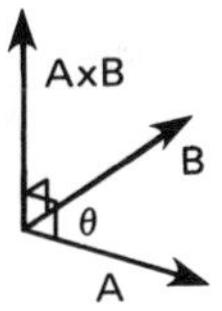

The *cross product* is a vector of magnitude $|\mathbf{B}||\mathbf{A}|\sin\theta$ which is perpendicular to the plane containing **A** and **B**.

$$\mathbf{A}\times\mathbf{B} = \begin{vmatrix} \mathbf{i} & \mathbf{j} & \mathbf{k} \\ A_x & A_y & A_z \\ B_x & B_y & B_z \end{vmatrix}$$

The sense of $\mathbf{A}\times\mathbf{B}$ is determined by the right-hand rule.

TRIGONOMETRY (6-2)

Trigonometric functions are defined in terms of a right triangle.

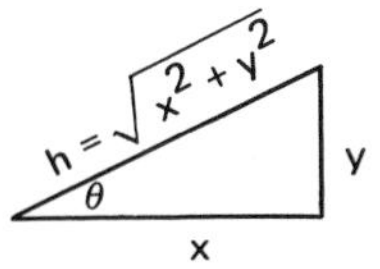

$$\sin\theta = \frac{y}{h} \qquad \cos\theta = \frac{x}{h} \qquad \tan\theta = \frac{y}{x}$$

$$\csc\theta = \frac{h}{y} \qquad \sec\theta = \frac{h}{x} \qquad \cot\theta = \frac{x}{y}$$

Basic Identities (6-3)

$$\sin^2\theta + \cos^2\theta = 1$$

$$\sin 2\theta = 2\sin\theta\cos\theta$$

$$1 + \tan^2\theta = \sec^2\theta$$

$$\cos 2\theta = \cos^2\theta - \sin^2\theta = 2\cos^2\theta - 1 = 1 - 2\sin^2\theta$$

For small angles,

$$\sin\theta = \tan\theta = \theta \quad [\theta \text{ in radians}]$$

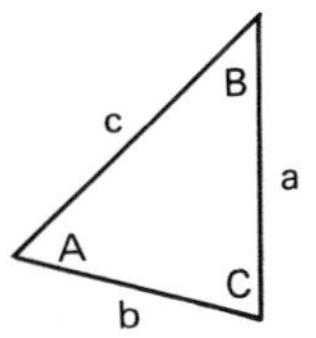

For general triangles, the *law of sines* is

$$\frac{\sin A}{a} = \frac{\sin B}{b} = \frac{\sin C}{c}$$

The *law of cosines* is

$$a^2 = b^2 + c^2 - 2bc\cos A$$

The area of a general triangle is $\frac{1}{2}ab(\sin C)$.

TWO-ANGLE FORMULAS (6-4)

$$\sin(\theta \pm \phi) = \sin\theta\cos\phi \pm \cos\theta\sin\phi$$

$$\cos(\theta \pm \phi) = \cos\theta\cos\phi \mp \sin\theta\sin\phi$$

$$\tan(\theta \pm \phi) = \frac{\tan\theta \pm \tan\phi}{1 \mp \tan\theta\tan\phi}$$

$$\cot(\theta \pm \phi) = \frac{\cot\phi\cot\theta \mp 1}{\cot\phi \pm \cot\theta}$$

HYPERBOLIC FUNCTIONS (6-4)

hyperbolic sine: $\sinh\theta = \dfrac{e^\theta - e^{-\theta}}{2}$

hyperbolic cosine: $\cosh\theta = \dfrac{e^\theta + e^{-\theta}}{2}$

hyperbolic tangent: $\tanh\theta = \dfrac{e^\theta - e^{-\theta}}{e^\theta + e^{-\theta}} = \dfrac{\sinh\theta}{\cosh\theta}$

hyperbolic cotangent: $\coth\theta = \dfrac{e^\theta + e^{-\theta}}{e^\theta - e^{-\theta}} = \dfrac{\cosh\theta}{\sinh\theta}$

hyperbolic secant: $\operatorname{sech}\theta = \dfrac{2}{e^\theta + e^{-\theta}} = \dfrac{1}{\cosh\theta}$

hyperbolic cosecant: $\operatorname{csch}\theta = \dfrac{2}{e^\theta - e^{-\theta}} = \dfrac{1}{\sinh\theta}$

HYPERBOLIC IDENTITIES (6-5)

$$\cosh^2\theta - \sinh^2\theta = 1$$

$$1 - \tanh^2\theta = \operatorname{sech}^2\theta$$

$$1 - \coth^2\theta = -\operatorname{csch}^2\theta$$

$$\cosh\theta + \sinh\theta = e^\theta$$

$$\cosh\theta - \sinh\theta = e^{-\theta}$$

$$\sinh(\theta \pm \phi) = \sinh\theta\cosh\phi \pm \cosh\theta\sinh\phi$$

$$\cosh(\theta \pm \phi) = \cosh\theta\cosh\phi \pm \sinh\theta\sinh\phi$$

$$\tanh(\theta \pm \phi) = \frac{\tanh\theta \pm \tanh\phi}{1 \pm \tanh\theta\tanh\phi}$$

COORDINATE SYSTEMS (7-3)

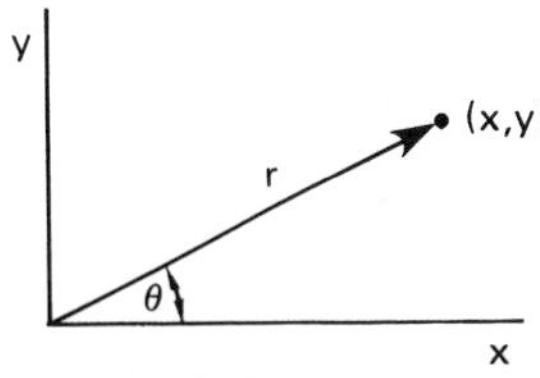

$x = r\cos\theta$
$y = r\sin\theta$

(a) polar

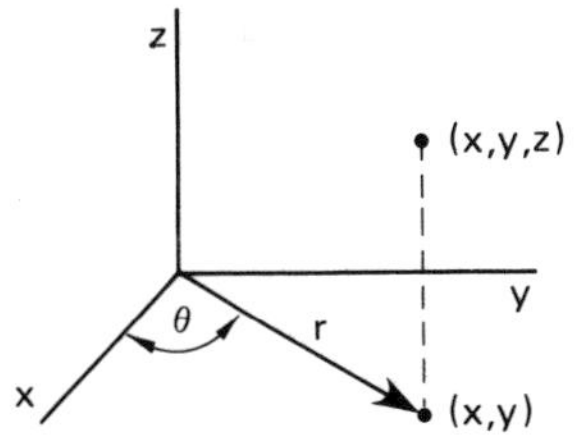

$x = r\cos\theta$
$y = r\sin\theta$
$z = z$

(b) cylindrical

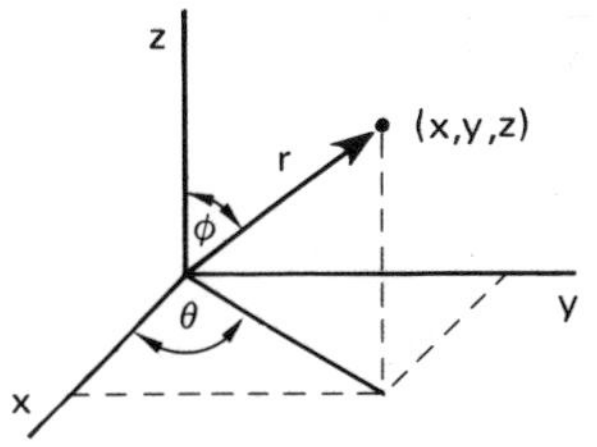

$x = r\sin\phi\cos\theta$
$y = r\sin\phi\sin\theta$
$z = r\cos\phi$

(c) spherical

EQUATIONS OF A STRAIGHT LINE (7-4)

The *general form* is $Ax + By + C = 0$.

The *slope-intercept form* is $y = mx + b$ where m is the slope, a is the x-intercept, and b is the y-intercept.

Intercept form: $\dfrac{x}{a} + \dfrac{y}{b} = 1$

Given one point, (x_1, y_1), on the line, the *point-slope form* is

$$y - y_1 = m(x - x_1)$$

MATHEMATICS

Slopes of perpendicular lines: $m_2 = \dfrac{-1}{m_1}$

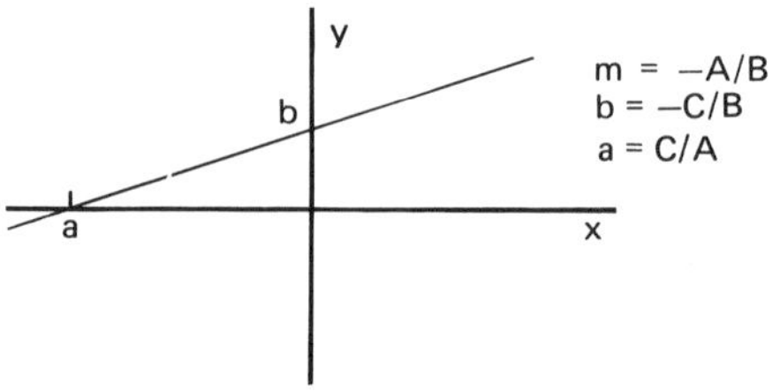

CONIC SECTIONS (7-8)

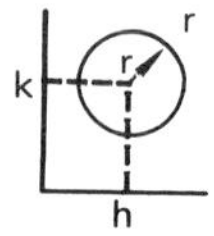

If the cutting plane is perpendicular to the axis of the cone, the intersection is a circle. The equation of a circle of radius r which is centered at (h, k) is

$$(x-h)^2 + (y-k)^2 = r^2$$

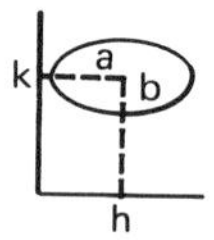

If the cutting plane is inclined to the axis of the cone, the intersection will be an ellipse. The equation of an ellipse centered at (h, k) with a and b defined as shown is

$$\left(\frac{x-h}{a}\right)^2 + \left(\frac{y-k}{b}\right)^2 = 1$$

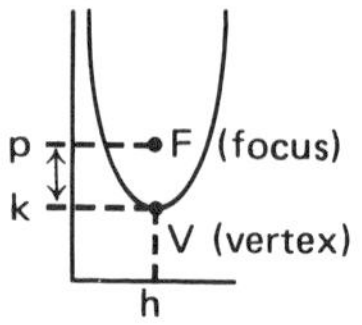

If the cutting plane is parallel to the cone surface, the intersection is a parabola. The equation of a parabola with vertex at (h, k) is

$$(x-h)^2 = 4p(y-k) \quad \text{[opens vertically]}$$

$$(y-k)^2 = 4p(x-h) \quad \text{[opens horizontally]}$$

MENSURATION (A-7) (App. 7.A)

Circle of radius r:

$$\text{circumference} = 2\pi r = p$$

$$\text{area} = \pi r^2 = \frac{p^2}{4\pi}$$

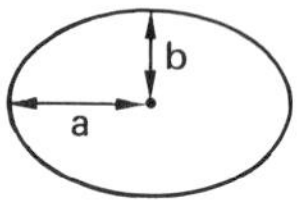

Ellipse with semimajor and semiminor distances a and b:

$$\text{area} = \pi ab$$

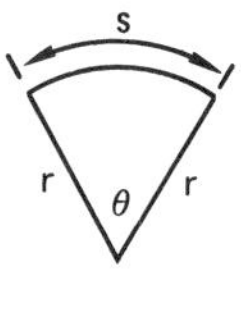

Circular Sector with radius r, included angle θ, and arc length s:

$$\text{area} = \tfrac{1}{2}\theta r^2 = \tfrac{1}{2}sr$$

$$\text{arc length} = s = \theta r$$

The angle θ is in radians.

VOLUME AND SURFACE AREAS (A-9) (App. 7.B)

A = cross-sectional area or end area

S = lateral surface area

V = enclosed volume

Right Circular Cylinder

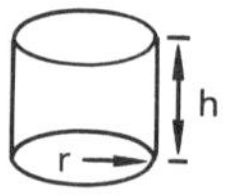

$$A = \pi r^2 = \tfrac{\pi}{4}d^2$$
$$S = 2\pi rh = \pi dh$$
$$V = Ah = \pi r^2 h = \tfrac{\pi}{4}d^2 h$$

Sphere

$$S = 4\pi r^2 = \pi d^2$$
$$V = \tfrac{4}{3}\pi r^3 = \tfrac{\pi}{6}d^3$$

Right Circular Cone

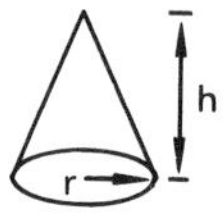

$$A_{\text{base}} = \pi r^2 = \tfrac{\pi}{4}d^2$$
$$S = \pi r\sqrt{r^2 + h^2}$$
$$V = \tfrac{1}{3}\pi r^2 h = \tfrac{\pi}{12}d^2 h$$

ELEMENTARY DERIVATIVE OPERATIONS (8-1)

$$\mathbf{D}k = 0 \qquad \mathbf{D}\cos x = -\sin x$$
$$\mathbf{D}x^n = nx^{n-1} \qquad \mathbf{D}\tan x = \sec^2 x$$
$$\mathbf{D}\ln x = \frac{1}{x} \qquad \mathbf{D}\cot x = -\csc^2 x$$
$$\mathbf{D}e^{ax} = ae^{ax} \qquad \mathbf{D}\sec x = \sec x \tan x$$
$$\mathbf{D}\sin x = \cos x \qquad \mathbf{D}\csc x = -\csc x \cot x$$

$$\mathbf{D}\arcsin x = \frac{1}{\sqrt{1-x^2}}$$

$$\mathbf{D}\arccos x = -\mathbf{D}\arcsin x$$

$$\mathbf{D}\arctan x = \frac{1}{1+x^2}$$

$$\mathbf{D}\,\mathrm{arccot}\, x = -\mathbf{D}\arctan x$$

$$\mathbf{D}\,\mathrm{arcsec}\, x = \frac{1}{x\sqrt{x^2-1}}$$

$$\mathbf{D}\,\mathrm{arccsc}\, x = -\mathbf{D}\,\mathrm{arcsec}\, x$$

$$\mathbf{D}\sinh x = \cosh x$$

$$\mathbf{D}\cosh x = \sinh x$$

$$\mathbf{D}\tanh x = \mathrm{sech}^2 x$$

$$\mathbf{D}\coth x = -\mathrm{csch}^2 x$$

$$\mathbf{D}\,\mathrm{sech}\, x = -\mathrm{sech}\, x \tanh x$$

$$\mathbf{D}\,\mathrm{csch}\, x = -\mathrm{csch}\, x \coth x$$

EXTREMA BY DIFFERENTIATION (8-2)

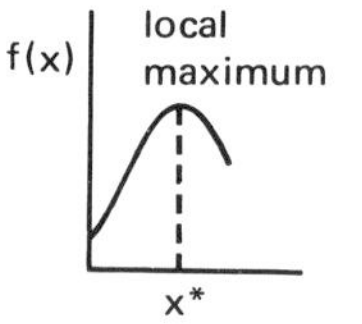

Given a continuous function, $f(x)$, the extreme points can be found by taking the first derivative and setting it equal to zero. Let x^* be the value of x which satisfies this equality. $f(x^*)$ is a minimum if $f''(x^*)$ is greater than zero. If $f''(x^*)$ is less than zero, $f(x^*)$ is a maximum. $f''(x^*)$ is equal to zero at an *inflection point.*

TAYLOR'S FORMULA (8-7)

$$f(b) = f(a) + \frac{f'(a)}{1!}(b-a) + \frac{f''(a)}{2!}(b-a)^2 + \cdots + \frac{f^n(a)}{n!}(b-a)^n + R_n(b)$$

COMMON SERIES APPROXIMATIONS (8-8)

Taylor's formulas can be used (by expanding about $a = 0$) to derive the following series approximations.

$$\sin x \approx x - \frac{x^3}{3!} + \frac{x^5}{5!} - \frac{x^7}{7!} + \cdots + (-1)^n \frac{x^{2n+1}}{(2n+1)!}$$

$$\cos x \approx 1 - \frac{x^2}{2!} + \frac{x^4}{4!} - \frac{x^6}{6!} + \cdots + (-1)^n \frac{x^{2n}}{(2n)!}$$

$$\sinh x \approx x + \frac{x^3}{3!} + \frac{x^5}{5!} + \frac{x^7}{7!} + \cdots + \frac{x^{2n+1}}{(2n+1)!}$$

$$\cosh x \approx 1 + \frac{x^2}{2!} + \frac{x^4}{4!} + \frac{x^6}{6!} + \cdots + \frac{x^{2n}}{(2n)!}$$

$$e^x \approx 1 + x + \frac{x^2}{2!} + \frac{x^3}{3!} + \cdots + \frac{x^n}{n!}$$

$$\ln(1+x) \approx x - \frac{x^2}{2} + \frac{x^3}{3} - \frac{x^4}{4} + \cdots + (-1)^{n+1}\frac{x^n}{n}$$

$$\frac{1}{1-x} \approx 1 + x + x^2 + x^3 + \cdots + x^n$$

ELEMENTARY INTEGRAL OPERATIONS (9-1)

$$\int k\,dx = kx + C$$

$$\int x^m\,dx = \frac{x^{m+1}}{m+1} + C \quad [m \neq -1]$$

$$\int \frac{1}{x} dx = \ln|x| + C$$

$$\int e^{kx}\,dx = \frac{e^{kx}}{k} + C$$

$$\int xe^{kx}\,dx = \frac{e^{kx}(kx-1)}{k^2} + C$$

$$\int k^{ax}\,dx = \frac{k^{ax}}{a \ln k} + C$$

$$\int \ln x\,dx = x \ln x - x + C$$

$$\int \sin x\,dx = -\cos x + C$$

$$\int \cos x\,dx = \sin x + C$$

$$\int \tan x\,dx = \ln \sec x + C$$

$$\int \cot x\,dx = \ln \sin x + C$$

$$\int \sec x\,dx = \ln(\sec x + \tan x) + C$$

$$\int \csc x\,dx = \ln(\csc x - \cot x) + C$$

$$\int \frac{dx}{k^2 + x^2} = \tfrac{1}{k} \arctan \tfrac{x}{k} + C$$

$$\int \frac{dx}{\sqrt{k^2 - x^2}} = \arcsin \tfrac{x}{k} + C \quad [k^2 > x^2]$$

$$\int \frac{dx}{x\sqrt{x^2 - k^2}} = \tfrac{1}{k}\operatorname{arcsec}\tfrac{x}{k} + C \quad [x^2 > k^2]$$

$$\int \sin^2 x\,dx = \tfrac{1}{2}x - \tfrac{1}{4}\sin 2x + C$$

$$\int \cos^2 x\,dx = \tfrac{1}{2}x + \tfrac{1}{4}\sin 2x + C$$

$$\int \tan^2 x\,dx = \tan x - x + C$$

AREA BETWEEN TWO CURVES (9-4)

$$A = \int_a^b (f_1(x) - f_2(x))dx$$

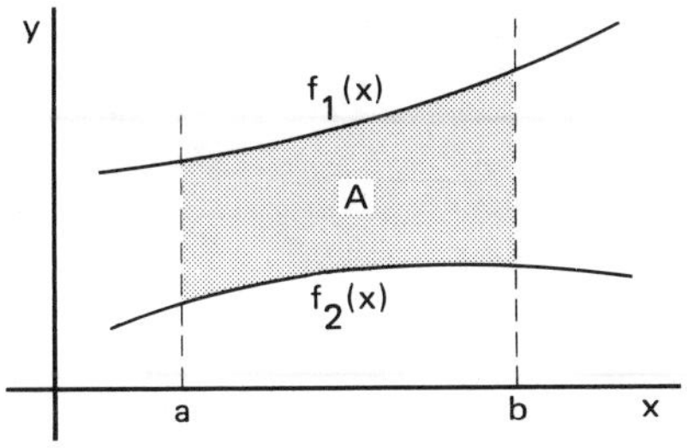

PERMUTATIONS AND COMBINATIONS (11-2)

$$P(n,r) = \frac{n!}{(n-r)!} \quad \text{[order conscious]}$$

$$C(n,r) = \frac{n!}{r!(n-r)!} \equiv \binom{h}{r} \quad \text{[not order conscious]}$$

BINOMIAL PROBABILITY DISTRIBUTION (11-4)

$p(x)$ is the probability that x will occur in n trials.
p is the probability of one success in one trial.
$q = (1-p)$ is the probability of one failure in one trial.

$$p(x) = \binom{n}{x} p^x q^{n-x} = \frac{n!}{(n-x)!x!} p^x q^{n-x}$$

POISSON PROBABILITY DISTRIBUTION (11-5)

$p(x)$ is the probability of x occurrences. x is the actual number of occurences in some period. λ is the mean number of occurences per period.

$$p(x) = \frac{e^{-\lambda}\lambda^x}{x!}$$

STATISTICS (11-9, 11-10)

$$\text{arithmetic mean} = \bar{x} = \left(\tfrac{1}{n}\right)(x_1 + x_2 + \cdots + x_n) = \frac{\sum x_i}{n}$$

$$\text{geometric mean} = \sqrt[n]{x_1 x_2 x_3 \cdots x_n}$$

$$\text{harmonic mean} = \frac{n}{\frac{1}{x_1} + \frac{1}{x_2} + \cdots + \frac{1}{x_n}}$$

$$\text{root-mean-squared value} = \sqrt{\frac{\sum x_i^2}{n}}$$

$$\text{standard deviation} = \sigma = \sqrt{\frac{\sum (x_i - \mu)^2}{N}} = \sqrt{\frac{\sum x_i^2}{N} - (\mu)^2}$$

$$\text{sample standard deviation} = s = \sqrt{\frac{\sum (x_i - \bar{x})^2}{n-1}}$$

$$= \sqrt{\frac{\sum x_i^2 - \frac{\left(\sum x_i\right)^2}{n}}{n-1}}$$

variance $= \sigma^2$
sample variance $= s^2$

ENGINEERING ECONOMICS

NOMENCLATURE AND DEFINITIONS

A	*Annual amount* or *annuity*: a payment or receipt of a fixed sum of money at yearly intervals.
	Amortization: the setting aside of money at intervals, as in a sinking fund, for gradual payment of a debt; or, the writing off of capital investments by prorating initial cost over a fixed period.
B	*Book value*: the theoretical market value of an asset at any time within its useful life. Book value is calculated as the initial cost less any accumulated depreciation.
C	*Cost*: the asset purchase price.
d	*Declining balance depreciation rate*: For double declining balance depreciation, $d = 2/n$.
D_j	*Depreciation* in year j: the allowance in year j of the life of the asset for the decrease in value due to wear, deterioration, and obsolescence.
EUAC	*Equivalent Uniform Annual Cost*: the present worth of an alternative multiplied by the (A/P) factor.
F	*Future worth*, value, or amount: the value of an asset at some future point in time. May be calculated by multiplying the present worth of an alternative by the (F/P) factor.
G	*Uniform gradient amount*: a quantity by which the cash flows change each year.
i	*Annual effective interest* rate.
k	Number of compounding periods per year.
MARR	*Minimum Attractive Rate of Return* (usually the same as the annual effective interest rate, i).
n	Number of compounding periods; or, the expected life of an asset.
P	*Present worth*, value, or amount. The value of an amount at time = 0.
	Nominal annual interest rate (same as rate per annum).
ROI	*Return On Investment*.
ROR	*Rate of Return*: the interest rate that makes the present worth equal to zero.
S_n	Expected *salvage value* in year n.
t	Time or *income tax* rate.
ϕ	*Effective interest rate per period* (equal to r/k).

YEAR-END CONVENTION (13-2)

All cash disbursements and receipts (cash flows) are assumed to occur at the end of the year in which they actually occur. The exception is an initial cost (purchase price) which is assumed to occur at time = 0.

SUNK COSTS (13-3)

Sunk costs are expenses incurred before time = 0. They have no bearing on the evaluation of alternatives.

CASH FLOW DIAGRAMS (13-3)

Cash flow diagrams are drawn to help visualize problems involving transfers of money at various points in time. The following conventions are used in their construction.

- The horizontal axis represents time. The axis is divided into equal increments, one per period.
- The year-end convention is assumed.
- Disbursements are downward arrows; receipts are upward arrows. In both cases, the length of the arrow is proportional to the magnitude of the transfer.
- Two or more transfers in the same time period are represented by head-to-tail arrows.

DISCOUNTING FACTORS (13-6)

Discounting factors are numbers used to calculate the equivalent amount (at some point in time) of an alternative. The factors are given the functional notation $(X/Y, i\%, n)$ where X is the desired value, Y is the known value, i is the interest rate, and n is the number of periods. (Usually, n is in years.) The functional symbol is read "...X given Y at $i\%$ for n years ...". The discounting factors convert Y to X under the conditions of constant interest rate compounded for n years. The common discounting factors (for discrete compounding) are

single payment compound amount	$(F/P, i\%, n)$	$(1+i)^n$
present worth	$(P/F, i\%, n)$	$(1+i)^{-n}$
uniform series sinking fund	$(A/F, i\%, n)$	$\dfrac{i}{(1+i)^n - 1}$
capital recovery	$(A/P, i\%, n)$	$\dfrac{i(1+i)^n}{(1+i)^n - 1}$
compound amount	$(F/A, i\%, n)$	$\dfrac{(1+i)^n - 1}{i}$
equal series present worth	$(P/A, i\%, n)$	$\dfrac{(1+i)^n - 1}{i(1+i)^n}$
uniform gradient	$(P/G, i\%, n)$	$\left(\dfrac{1}{i} - \dfrac{n}{(1+i)^n - 1}\right)(P/A, i\%, n)$

RATE OF RETURN (13-11)

The *rate of return* (ROR) is the interest rate that makes the present worth equal to zero. To determine the ROR, assume a reasonable value for the interest rate, i, and find the present worth. If the present worth is zero, the assumed interest rate is the ROR. If the present worth is not zero, assume another value of i and find the present worth. Use the two values of i and their corresponding present worths to interpolate or extrapolate a value of i that makes the present worth zero.

COMPARING ALTERNATIVES (13-14, 13-15)

The *present worth* method may be used if all of the alternatives have equal lives. The present worth of each alternative is calculated. The alternative with the smallest negative or largest positive present worth is chosen.

The *equivalent uniform annual cost* (EUAC) method must be used if alternatives have unequal lives. Use of this method is restricted to alternatives which are renewed—specifically, each alternative is replaced by an identical replacement at the end of its useful life, infinitely or up to the duration of the longest-lived alternative. The EUAC of an alternative is

$$\text{EUAC} = \begin{pmatrix}\text{present worth}\\ \text{of alternative}\end{pmatrix}(A/P, i\%, n)$$

The *capitalized cost* of a project is the present worth of a project that has an infinite life. The capitalized cost represents the amount of money needed at time = 0 to support the project forever on interest only, without reducing the principal. For annual operating and maintenance costs,

$$\begin{matrix}\text{capitalized}\\ \text{cost}\end{matrix} = \begin{matrix}\text{initial}\\ \text{cost}\end{matrix} + \frac{\text{annual maintenance cost}}{i}$$

For a project with maintenance costs every n years,

$$\begin{matrix}\text{capitalized}\\ \text{cost}\end{matrix} = \begin{matrix}\text{initial}\\ \text{cost}\end{matrix} + \frac{(\text{maintenance cost})(A/P, i\%, n)}{i}$$

The *benefit/cost ratio* is

$$B/C = \frac{\begin{matrix}\text{present worth}\\ \text{of benefits}\end{matrix} - \begin{matrix}\text{present worth}\\ \text{of disbenefits}\end{matrix}}{\text{present worth of costs}}$$

DEPRECIATION (13-20)

straight line $\quad D_j = \dfrac{C - S_n}{n}$

double declining balance ($i = 1$ to $j - 1$) $\quad D_j = \dfrac{2C}{n}\left(1 - \dfrac{2}{n}\right)^{j-1}$

$$\text{BV}_j = C\left(1 - \frac{2}{n}\right)^j$$

sum of the years' digits $\quad D_j = \dfrac{(C - S_n)(n - j + 1)}{T}$

$$T = \tfrac{1}{2}n(n+1)$$

sinking fund $\quad D_j = (C - S_n)(A/F, i\%, n) \times (F/P, i\%, j - 1)$

ACRS or MACRS $\quad D_j = (\text{factor})C$

BOOK VALUE (13-22)

$$\text{BV} = \text{initial cost} - \sum D_j$$

INCOME TAXES (13-24)

If income taxes are paid, operating expenses and depreciation are deductible. If t is the tax rate, revenues and all expenses (except depreciation) should be multiplied by $(1-t)$ in the year in which they occur. Although depreciation is a deductible expense, it is not an actual out-of-pocket expense. Depreciation should be multiplied by t and added to the cash flow in the appropriate year.

$$t = s + f - sf$$

NONANNUAL COMPOUNDING (13-27)

For problems in which compounding is done at intervals other than yearly, an effective annual interest rate can be computed from

$$i = \left(1 + \frac{r}{k}\right)^k - 1 = (1 + \phi)^k - 1$$

$$\phi = \frac{r}{k} = \text{effective rate per period}$$

HANDLING INFLATION (13-37)

It is important to perform economic studies in terms of *constant value dollars*. One method of converting all cash flows to constant value dollars is to divide the flows by some annual *economic indicator* or *price index*.

An alternative is to replace i with a value corrected for the inflation rate, e. This corrected value, i', is

$$i' = i + e + ie$$

CONSUMER LOANS (13-38)

BAL_j	balance after the jth payment
LV	total value loaned (cost − down payment)
j	payment or period number
N	total number of payments to pay off the loan
PI_j	jth interest payment
PP_j	jth principal payment
PT_j	jth total payment
ϕ	effective rate per period (r/k)

SIMPLE INTEREST (13-38)

Interest due does not compound with a *simple interest* loan. The interest due is proportional to the length of time the principal is outstanding.

DIRECT REDUCTION LOANS (13-39)

This is the typical "interest paid on unpaid balance" loan. The amount of the periodic payment is constant, but the amounts paid toward the principal and interest both vary.

$$N = \frac{-\ln\left(1 - \dfrac{-\phi(\text{LV})}{\text{PT}}\right)}{\ln(1+\phi)}$$

$$\text{PT} = \text{LV}(A/P, \phi\%, n)$$

$$\text{LV} = \frac{\text{PT}}{(A/P, \phi\%, n)}$$

$$\text{BAL}_{j-1} = \text{PT}\left(\frac{1 - (1+\phi)^{j-1-N}}{\phi}\right)$$

$$\text{PI}_j = \phi(\text{BAL}_{j-1})$$

$$\text{PP}_j = \text{PT} - \text{PI}_j$$

$$\text{BAL}_j = \text{BAL}_{j-1} - \text{PP}_j$$

FLUID STATICS AND DYNAMICS

ATMOSPHERIC PRESSURE (14-3)

1.000 atm	760.0 torr
14.696 psia	1.013 bars
407.1 in w.g.	1013 millibars
33.93 ft w.g.	1.013×10^5 Pa
29.921 in Hg	101.3 kPa
760.0 mm Hg	

DENSITY AND SPECIFIC GRAVITY (14-3)

water: 62.4 lbm/ft^3, 0.0361 lbm/in^3, 8.345 lbm/gal; SG = 1.0
mercury: 848.4 lbm/ft^3, 0.491 lbm/in^3; SG = 13.6
air: 0.075 lbm/ft^3 at 1 atm and 70°F

VISCOSITY (14-6)

μ: absolute (dynamic) viscosity (lbf-sec/ft^2)

$$= \frac{\text{shear stress}}{\text{rate of shear strain}}$$

ν: kinematic viscosity (ft^2/sec) $= \dfrac{\mu g_c}{\rho}$

SPEED OF SOUND IN A FLUID (14-13)

$$a = \sqrt{kg_cRT} = \sqrt{\frac{kg_cp}{\rho}} \quad \text{[gases]}$$

$$a = \sqrt{\frac{Eg_c}{\rho}} \quad \text{[liquids]}$$

$$E \text{ (bulk modulus)} = \frac{1}{\text{compressibility}}$$

$$\approx 4.3 \times 10^7 \text{ lbf/ft}^2 \text{ for water}$$

$$M \text{ (Mach number)} = \frac{\text{v}}{a}$$

MANOMETERS (15-3)

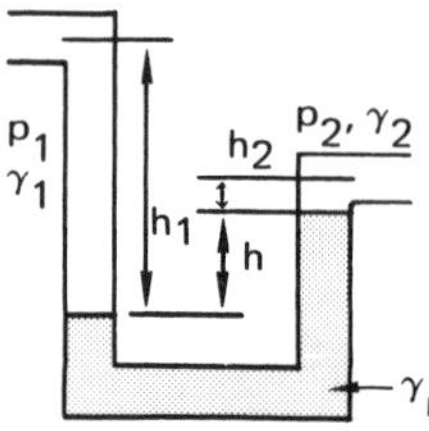

$$\Delta p = p_1 - p_2$$

$$= \gamma_m h - \gamma_1 h_1 + \gamma_2 h_2$$

HYDROSTATIC PRESSURE ON A FLAT PLANE (15-6)

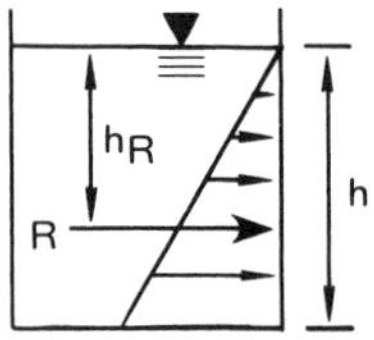

$$\gamma = \rho \times \frac{g}{g_c}$$

$$p = \gamma h$$

$$\bar{p} = \tfrac{1}{2}\gamma h$$

$$R = \bar{p}A = \tfrac{1}{2}\gamma hA$$

$$h_R = \tfrac{2}{3}h$$

PRESSURE ON SUBMERGED PLANE SURFACES (15-6)

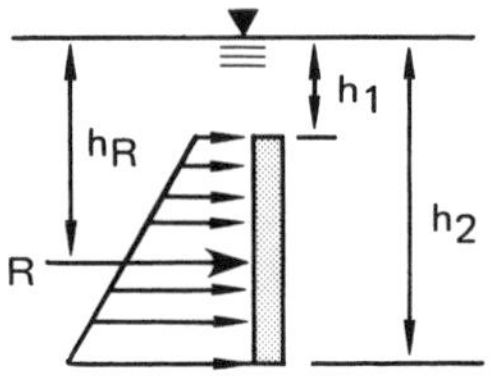

h_c = distance from surface to centroid of plane, as measured parallel to plane's surface

$$\bar{p} = \gamma h_c = \tfrac{1}{2}\gamma(h_1 + h_2)$$

$$R = \bar{p}A$$

$$h_R = h_c + \frac{I_c}{Ah_c} \quad \text{[as measured parallel to plane's surface]}$$

PRESSURE ON INCLINED PLANE SURFACE (15-7)

$$\bar{p} = \tfrac{1}{2}(p_1 + p_2) = \frac{\tfrac{1}{2}\rho g(h_1 + h_2)}{g_c}$$

$$= \frac{\tfrac{1}{2}\rho g(h_3 + h_4)\sin\theta}{g_c}$$

$$R = \bar{p}A$$

$$h_R = \frac{\tfrac{2}{3}}{\sin\theta}\left(h_1 + h_2 - \frac{h_1h_2}{h_1 + h_2}\right)$$

$$= \tfrac{2}{3}\left(h_3 + h_4 - \frac{h_3h_4}{h_3 + h_4}\right)$$

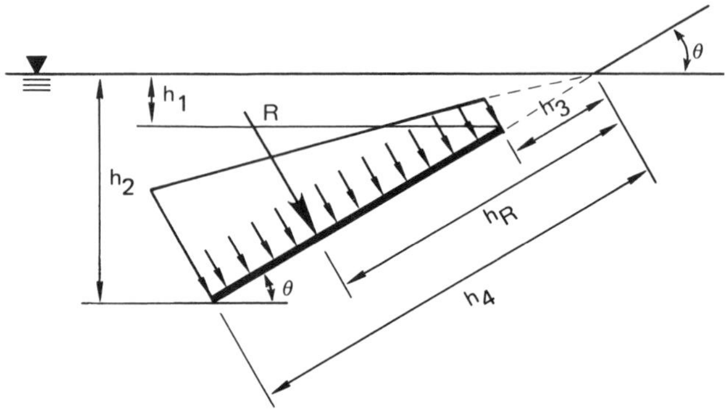

PRESSURE ON GENERAL PLANE SURFACE (15-8)

$$\bar{p} = \frac{\rho g h_c \sin\theta}{g_c} = \gamma h_c \sin\theta$$

$$R = \bar{p}A$$

$$h_R = h_c + \frac{I_c}{Ah_c}$$

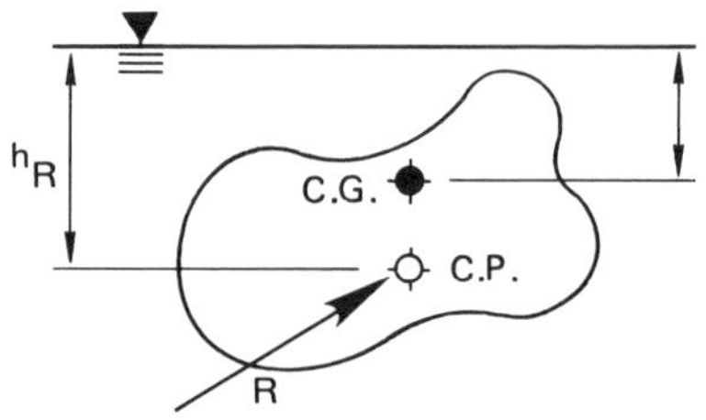

BUOYANCY (ARCHIMEDES' PRINCIPLE) (15-14)

The buoyant force is equal to the weight of the displaced fluid.

$$F_{\text{buoyant}} = \gamma V_{\text{displaced}}$$

$$V_{\text{displaced}} = \frac{\text{dry weight} - \text{submerged weight}}{\gamma}$$

BERNOULLI EQUATION (16-2)

$$\frac{p}{\rho} + \frac{\mathrm{v}^2}{2g_c} + \frac{zg}{g_c} = \text{constant}$$

$$\frac{p}{\rho} = \text{pressure (static) head}$$

$$\frac{\mathrm{v}^2}{2g_c} = \text{velocity (dynamic) head}$$

$$\frac{zg}{g_c} = \text{potential (gravitational) head}$$

REYNOLDS NUMBER (16-7)

$$\mathrm{Re} = \frac{\mathrm{v}D\rho}{\mu g_c} = \frac{\mathrm{v}D}{\nu} = \frac{\text{inertial force}}{\text{viscous force}} \quad \text{[dimensionless]}$$

$$D = 4r_h \quad \text{[equivalent diameter]}$$

laminar $2100 > \mathrm{Re} > 4000$ turbulent

CONTINUITY EQUATION (17-2)

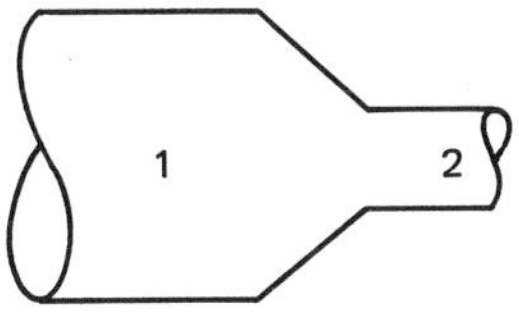

$$\dot{m}_1 = \dot{m}_2$$

$$A_1\mathrm{v}_1\rho_1 = A_2\mathrm{v}_2\rho_2 \quad \text{[gases and liquids]}$$

$$A_1\mathrm{v}_1 = A_2\mathrm{v}_2 \quad \text{[incompressible liquids only]}$$

$$A\mathrm{v}\rho = \dot{m}$$

$$A\mathrm{v} = \dot{V}$$

FRICTION LOSS (DARCY) (17-6)

$$h_f = \frac{fL\mathrm{v}^2}{2Dg} = \frac{K\mathrm{v}^2}{2g} = E_f \times \frac{g_c}{g}$$

K is the friction loss coefficient for fittings.

$$f = \frac{64}{\mathrm{Re}} \text{ for } \mathrm{Re} < 2100 \quad \text{[laminar flow only]}$$

MINOR LOSSES (17-9)

$$h_m = Kh_\mathrm{v} = K\left(\frac{\mathrm{v}^2}{2g}\right) = \frac{fL_e\mathrm{v}^2}{2Dg}$$

$$K = \frac{fL_e}{D}$$

CONSERVATION OF ENERGY (17-11)

$$\frac{p_1}{\rho} + \frac{\mathrm{v}_1^2}{2g_c} + \frac{z_1 g}{g_c} + E_\text{pump} = \frac{p_2}{\rho} + \frac{\mathrm{v}_2^2}{2g_c} + \frac{z_2 g}{g_c} + E_f + E_\text{turbine}$$

E_pump, E_f, and E_turbine are on a per-pound basis with units of ft-lbf/lbm.

TORRICELLI EQUATION (17-13)

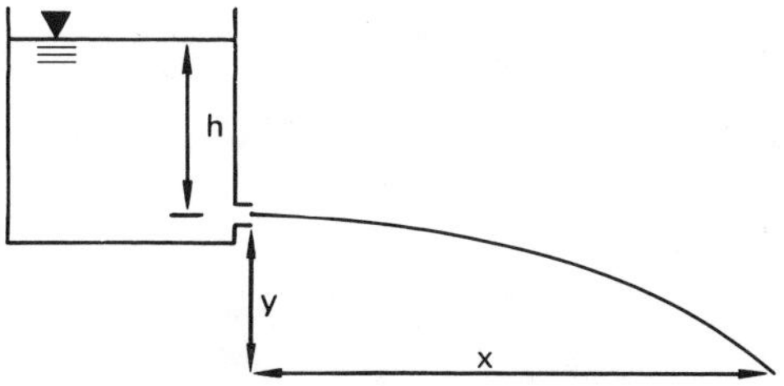

$$C_\mathrm{v} = \frac{C_d}{C_c}$$

$$\mathrm{v}_o = C_\mathrm{v}\sqrt{2gh}$$

$$y = \tfrac{1}{2}gt^2$$

$$x = \mathrm{v}_o t$$

$$\dot{V} = C_d A_o \mathrm{v}_o$$

STATIC-PITOT TUBE (17-18)

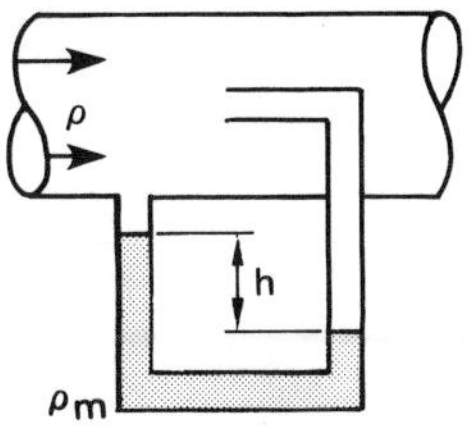

$$\mathrm{v} = \sqrt{2g\frac{(\rho_m - \rho)h}{\rho}}$$

$\rho \approx 0$ for air

VENTURI METER (17-19)

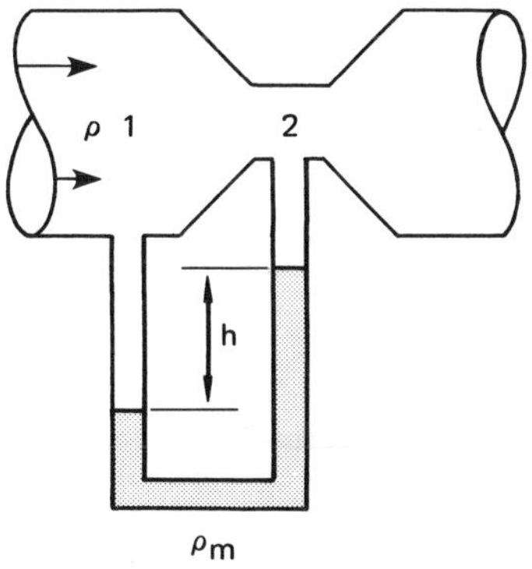

$$\beta = \frac{D_2}{D_1} = \sqrt{\frac{A_2}{A_1}}$$

$$F_\text{va} = \frac{1}{\sqrt{1-\beta^4}}$$

$$\mathrm{v}_2 = C_\mathrm{v}F_\text{va}\sqrt{\frac{2gh(\rho_m - \rho)}{\rho}}$$

$$= C_\mathrm{v}F_\text{va}\sqrt{\frac{2g_c(p_1 - p_2)}{\rho}}$$

$$C_d = C_c C_\mathrm{v}$$

$$C_c \approx 1 \quad \text{[venturi meters]}$$

$$C_f = C_d F_\text{va}$$

$$\dot{V} = C_f A_2\sqrt{2gh\frac{(\rho_m - \rho)}{\rho}}$$

$$\rho \approx 0 \quad \text{[air]}$$

ORIFICE PLATE (17-20)

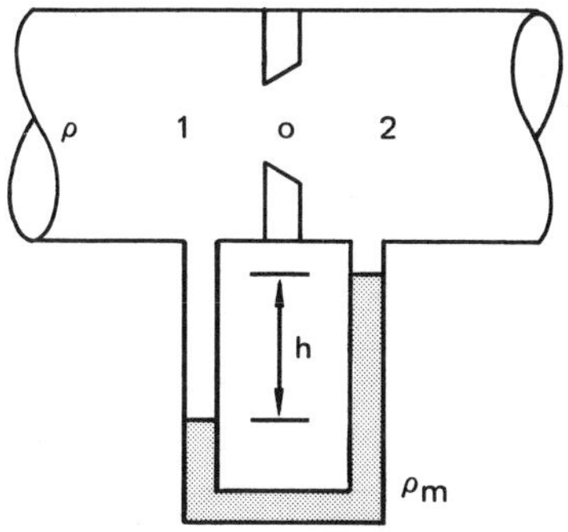

$$F_\text{va} = \frac{1}{\sqrt{1 - C_c^2\left(\frac{A_o}{A_1}\right)^2}}$$

$$\mathrm{v}_2 = F_\text{va}C_\mathrm{v}\sqrt{2gh\frac{(\rho_m - \rho)}{\rho}}$$

$$C_f = F_\text{va}C_d$$

$$C_d = C_c C_\mathrm{v}$$

$$\dot{V} = C_c A_o \mathrm{v}_2$$

$$= A_o C_f\sqrt{2gh\frac{(\rho_m - \rho)}{\rho}}$$

IMPULSE-MOMENTUM (17-22)

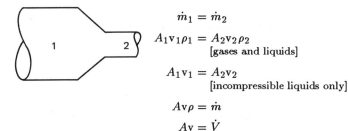

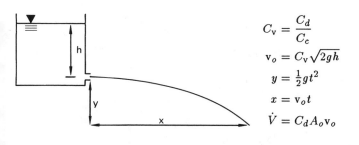

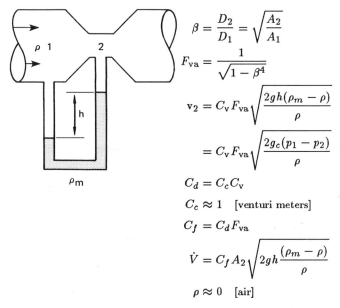

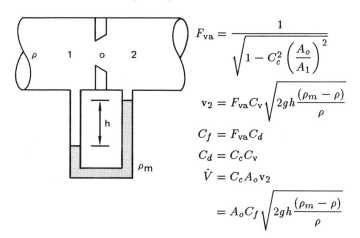

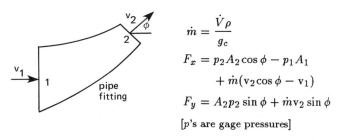

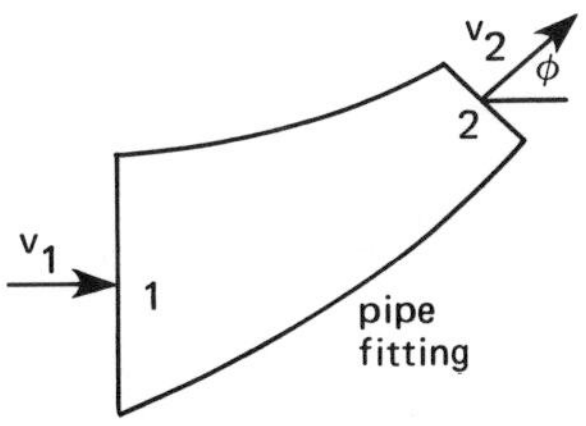

$$\dot{m} = \frac{\dot{V}\rho}{g_c}$$

$$F_x = p_2A_2\cos\phi - p_1A_1 + \dot{m}(\mathrm{v}_2\cos\phi - \mathrm{v}_1)$$

$$F_y = A_2p_2\sin\phi + \dot{m}\mathrm{v}_2\sin\phi$$

[p's are gage pressures]

FLUID STATICS AND DYNAMICS

LIFT AND DRAG (17-28, 17-29)

A is the projected area (a circle for a sphere).

$$F_L = \frac{C_L v^2 \rho A}{2g_c}$$

$$F_D = \frac{C_D v^2 \rho A}{2g_c}$$

SIMILARITY (17-32)

For fans, pumps, turbines, drainage through holes in tanks, closed-pipe flow with no free surfaces (in the turbulent region with the same relative roughness), and for completely submerged objects such as torpedoes, airfoils, and submarines, performance of models and prototypes can be correlated by equating Reynolds numbers.

$$(\text{Re})_{\text{model}} = (\text{Re})_{\text{prototype}}$$

$$\text{Re} = \frac{L v}{\nu} = \frac{L v \rho}{\mu g_c}$$

HYDRAULIC HORSEPOWER (W_{pump}) (18-4)

	Q in gpm	$\dot{m}$ in lbm/sec	$\dot{V}$ in cfs
h_A is added head in feet	$\frac{h_A Q(\text{SG})}{3956}$	$\frac{h_A \dot{m}}{550} \times \frac{g}{g_c}$	$\frac{h_A \dot{V}(\text{SG})}{8.814}$
Δp is added head in psf	$\frac{\Delta p Q}{2.468 \times 10^5}$	$\frac{\Delta p \dot{m}}{(34{,}320)(\text{SG})} \times \frac{g}{g_c}$	$\frac{\Delta p \dot{V}}{550}$

PUMP AFFINITY/SIMILARITY LAWS (18-11)

$$\frac{Q_2}{Q_1} = \frac{n_2}{n_1}$$

$$\frac{h_2}{h_1} = \left(\frac{n_2}{n_1}\right)^2 = \left(\frac{Q_2}{Q_1}\right)^2$$

$$\frac{P_2}{P_1} = \left(\frac{n_2}{n_1}\right)^3 = \left(\frac{Q_2}{Q_1}\right)^3$$

$$\frac{Q_2}{Q_1} = \frac{D_2}{D_1}$$

$$\frac{h_2}{h_1} = \left(\frac{D_2}{D_1}\right)^2$$

$$\frac{P_2}{P_1} = \left(\frac{D_2}{D_1}\right)^3$$

OPEN CHANNEL FLOW (Chap. 19)

$$r_h = \text{hydraulic radius} = \frac{\text{area in flow}}{\text{wetted perimeter}} = \frac{D_e}{4}$$

Chezy Equation (19-3)

$$v = C\sqrt{r_h S}$$

$$S = \text{slope} = \frac{h_f}{L}$$

For uniform flow, S = geometric slope.

Manning Equation (19-4) (omit 1.486 if r_h is in meters)

$$C = \frac{(1.486)(r_h)^{1/6}}{n}$$

n = Manning constant

Diameter of a Pipe Flowing Full (19-5)

$$D = 1.335\left(\frac{Qn}{\sqrt{S}}\right)^{3/8}$$

Diameter of a Pipe Flowing Half-Full (19-5)

$$D = 1.731\left(\frac{Qn}{\sqrt{S}}\right)^{3/8}$$

Hazen-Williams Equation

$$v = 1.318C(r_h)^{0.63}(S)^{0.54}$$

C = Hazen-Williams coefficient

Froude Number (19-13)

$$N_{\text{Fr}} = \frac{v}{\sqrt{g \times \text{depth}}}$$

PROPERTIES OF WATER AT ROOM TEMPERATURE

specific gravity (SG):	1
density (ρ):	62.4 lbm/ft^3 0.0361 lbm/in^3 8.345 lbm/gal 1000 kg/m^3 1.94 slug/ft^3
specific weight (γ):	62.4 lbf/ft^3 9800 N/m^3
absolute viscosity (μ):	2×10^{-5} lbf-sec/ft^2 1×10^{-3} Pa·s
kinematic viscosity (ν):	1×10^{-5} ft^2/sec 1×10^{-6} m^2/s
bulk modulus (E):	3.2×10^5 lbf/in

PROPERTIES OF STANDARD AIR

density (ρ):	0.075 lbm/ft^3 0.0023 slug/ft^3 1.23 kg/m^3
specific weight (γ):	0.075 lbf/ft^3 11.9 N/m^3
absolute viscosity (μ):	3.8×10^{-7} lbf-sec/ft^2 1.8×10^{-5} Pa·s
kinematic viscosity (ν):	1.6×10^{-4} ft^2/sec 1.6×10^{-5} m^2/s

PROPERTIES OF MERCURY

specific gravity (SG):	13.6
density (ρ):	848.4 lbm/ft^3 0.491 lbm/in^3 13,600 kg/m^3

IMPORTANT CONVERSIONS (A-1)

$$1 \text{ BTU} = 778 \text{ ft-lbf} \qquad \left(J = 778 \,\frac{\text{ft-lbf}}{\text{BTU}}\right)$$

$$1 \text{ hp} = 550 \text{ ft-lbf/sec} = 2545 \text{ BTU/hr}$$

$$1 \text{ kW} = 3412.9 \text{ BTU/hr}$$

$$1 \text{ kW} = 1.341 \text{ hp}$$

STP (STANDARD TEMPERATURE AND PRESSURE) (14-4)

Scientific STP is 32°F and 14.7 psia.
For fuel gases, STP is 60°F and 14.7 psia.

THE ZEROTH LAW OF THERMODYNAMICS (21-4)

If two bodies are in thermal equilibrium with a third body, then the two bodies are in thermal equilibrium with each other.

THE FIRST LAW OF THERMODYNAMICS (22-4)

One form of energy can be converted into another form, but energy cannot be created or destroyed.

MATHEMATICAL FORMULATION OF THE FIRST LAW (22-4)

For closed systems,

$$dQ = dU + W \quad \text{or} \quad Q = \Delta U + W$$

For steady-flow open systems,

$$q = \frac{W}{J} + h_2 - h_1 + \frac{(z_2 - z_1)g}{g_c J} + \frac{\text{v}_2^2 - \text{v}_1^2}{2g_c J}$$

[q, W, and h are on a per-pound basis]

FIRST LAW SIGN CONVENTION (22-4)

Heat (Q) is positive when it enters a system. Heat is negative when it leaves a system. Work (W) is positive when the system does work on the surroundings. Work is negative when the surroundings do work on the system.

THE SECOND LAW OF THERMODYNAMICS (22-11)

Heat will not flow by itself from a cold body to a hot body. Entropy always increases.

THE THIRD LAW OF THERMODYNAMICS (21-6)

The entropy of a perfect substance (crystal) at absolute zero is zero.

TEMPERATURE SCALES (21-4)

$$°\text{R} = °\text{F} + 460$$

$$°\text{K} = °\text{C} + 273$$

$$°\text{F} = 32 + (\tfrac{9}{5})°\text{C}$$

$$°\text{C} = (\tfrac{5}{9})(°\text{F} - 32)$$

$$\Delta°\text{R} = \Delta°\text{F} = (\tfrac{9}{5})\Delta°\text{K} = (\tfrac{9}{5})\Delta°\text{C}$$

$$\Delta°\text{K} = \Delta°\text{C} = (\tfrac{5}{9})\Delta°\text{R} = (\tfrac{5}{9})\Delta°\text{F}$$

PROPERTIES OF WATER (21-7)

$$c_p = 1.0 \text{ BTU/lbm-°F} = 4.19 \text{ kJ/kg·K}$$

SENSIBLE HEAT (21-8)

$$Q = mc_p(T_2 - T_1)$$

$$c_p = 1 \text{ BTU/lbm-°F} = 1 \text{ cal/g·°C} = 4.19 \text{ kJ/kg·K}$$

[for water]

LATENT HEAT (21-8)

Latent heat is the energy that changes the phase of a substance.
Latent heat of fusion: 143.4 BTU/lbm for turning ice to liquid
Latent heat of vaporization: 970.3 BTU/lbm for turning liquid to steam
Latent heat of sublimation: 1220 BTU/lbm for turning ice to steam directly
The above values are for water phases at 1 atm.

PROPERTIES OF ICE (21-8)

heat of fusion:	143 BTU/lbm 334 kJ/kg
heat of sublimation:	1220 BTU/lbm 2840 kJ/kg

QUALITY OF A LIQUID-VAPOR MIXTURE (21-9)

$$x = \frac{\text{mass of vapor}}{\text{mass of liquid} + \text{mass of vapor}}$$

PROPERTIES OF A LIQUID-VAPOR MIXTURE (21-12)

$$h = h_f + xh_{fg}$$

$$u = u_f + xu_{fg}$$

$$v = v_f + xv_{fg}$$

$$s = s_f + xs_{fg}$$

EQUATIONS OF STATE FOR IDEAL GASES (21-12)

$$pV = nR^*T \quad [\text{for } n \text{ moles}]$$

$$pV = mRT \quad [\text{for } m \text{ pounds mass}]$$

$$R = \frac{R^*}{(\text{MW})}$$

UNIVERSAL GAS CONSTANT (21-13)

Units in SI and Metric Systems

8.3143 kJ/kmol·K
8314.3 J/kmol·K
0.08206 atm·l/mol·K
1.986 cal/mol·K
8.314 J/mol·K
82.06 atm·cm^3/mol·K
0.08206 atm·m^3/kmol·K
8314.3 kg·m^2/s^2·kmol·K
8314.3 m^3·Pa/kmol·K
8.314 × 10^7 erg/mol·K

Units in English Systems

1545.33 ft-lbf/lbmole-°R
1.986 BTU/lbmole-°R
0.7302 atm-ft^3/lbmole-°R
10.73 ft^3-lbf/in^2-lbmole-°R

PROPERTIES OF AIR (21-14)

$$c_p = 0.240 \text{ BTU/lbm-°F} = 1.0 \text{ kJ/kg·K}$$

$$c_v = 0.171 \text{ BTU/lbm-°F} = 0.72 \text{ kJ/kg·K}$$

$$R = 53.3 \text{ ft-lbf/lbm-°R} = 287 \text{ J/kg·K}$$

$$k = \frac{c_p}{c_v} = 1.4$$

MISCELLANEOUS GAS LAWS

Avogadro (21-13): Equal volumes of different gases under the same conditions of temperature and pressure contain the same number of molecules (6.022×10^{23} molecules per gram-mole).

Dalton (21-16): The total pressure of a mixture of gases is equal to the sum of the partial pressures. (The partial pressures are determined at the final mixture temperature and volume.)

Amagat (21-16): The total volume of a mixture of gases is equal to the sum of the partial volumes. (The partial volumes are determined at the final mixture temperature and pressure.)

SPECIFIC HEAT RELATIONSHIPS FOR IDEAL GASES (21-15)

$$Q = mc\Delta T$$

$$k = \frac{c_p}{c_v}$$

$$c_p - c_v = \frac{R}{J}; \qquad C_p - C_v = \frac{R^*}{J}$$

$$c_p = \frac{Rk}{J(k-1)}; \qquad C_p = \frac{R^* k}{J(k-1)}$$

MIXTURES OF IDEAL GASES (21-17)

Volumetrically weighted (same as *mole fraction weighted*): density, molecular weight of the mixture, and all molar properties

Gravimetrically weighted (same as *mass weighted*): specific internal energy, specific enthalpy, specific entropy, specific heats, and specific gas constant of the mixture

COMPRESSED GAS EQUATION OF STATE (21-18)

$$pV = mZRT$$

$$Z = \text{compressibility factor}$$

TYPES OF THERMODYNAMIC SYSTEMS (22-1)

A *system* is a region with artificially-chosen boundaries. A *closed system* is one in which matter does not cross the system boundaries. Energy may cross the system boundaries, however. Both energy and matter may cross the boundaries of an *open system*.

TYPES OF PROCESSES (22-2)

Isochoric (Isometric): constant volume
Isobaric: constant pressure
Isothermal: constant temperature
Polytropic: any process for which $p(V)^n$ is constant
Adiabatic: a process with no heat transfer ($q = 0$)
Isentropic: an adiabatic process for which $\Delta s = 0$
Throttling: an adiabatic process for which $\Delta h = 0$

IDEAL GAS PROCESS LAW (22-7)

$$\frac{p_1 V_1}{T_1} = \frac{p_2 V_2}{T_2}$$

THERMODYNAMIC RELATIONSHIPS FOR ANY PROCESS (IDEAL GASES) (22-7)

$$\Delta u = c_v \Delta T \qquad (c_v = 0.171 \text{ BTU/lbm-°F for air})$$

$$\Delta h = c_p \Delta T \qquad (c_p = 0.240 \text{ BTU/lbm-°F for air})$$

CONSTANT PRESSURE PROCESSES (CLOSED SYSTEM, IDEAL GAS) (22-7)

$$p_2 = p_1$$

$$T_2 = \frac{T_1 v_2}{v_1}$$

$$v_2 = \frac{v_1 T_2}{T_1}$$

$$q = h_2 - h_1 = c_p(T_2 - T_1) = c_v(T_2 - T_1) + \frac{p(v_2 - v_1)}{J}$$

$$u_2 - u_1 = c_v(T_2 - T_1) = \frac{p(v_2 - v_1)}{J(k-1)} = \frac{c_v p(v_2 - v_1)}{JR}$$

$$W = \frac{p(v_2 - v_1)}{J} = \frac{R(T_2 - T_1)}{J}$$

$$s_2 - s_1 = c_p \ln \frac{T_2}{T_1} = c_p \ln \frac{v_2}{v_1}$$

$$h_2 - h_1 = q = c_p(T_2 - T_1) = \frac{kp(v_2 - v_1)}{J(k-1)}$$

CONSTANT VOLUME PROCESSES (CLOSED SYSTEM, IDEAL GAS) (22-8)

$$p_2 = \frac{p_1 T_2}{T_1}$$

$$T_2 = \frac{T_1 p_2}{p_1}$$

$$v_2 = v_1$$

$$q = u_2 - u_1 = c_v(T_2 - T_1)$$

$$u_2 - u_1 = q = c_v(T_2 - T_1) = \frac{v(p_2 - p_1)}{J(k-1)}$$

$$W = 0$$

$$s_2 - s_1 = c_v \ln \left(\frac{T_2}{T_1}\right) = c_v \ln \left(\frac{p_2}{p_1}\right)$$

$$h_2 - h_1 = c_p(T_2 - T_1) = \frac{kv(p_2 - p_1)}{J(k-1)}$$

CONSTANT TEMPERATURE PROCESSES (CLOSED SYSTEM, IDEAL GAS) (22-8)

$$p_2 = \frac{p_1 v_1}{v_2}$$

$$T_2 = T_1$$

$$v_2 = \frac{v_1 p_1}{p_2}$$

$$q = W = \frac{p_1 v_1}{J} \ln \frac{v_2}{v_1} = T(s_2 - s_1)$$

$$u_2 - u_1 = 0$$

$$W = q = \frac{p_1 v_1}{J} \ln \frac{v_2}{v_1} = \frac{RT}{J} \ln \frac{p_1}{p_2}$$

$$s_2 - s_1 = \frac{q}{T} = \frac{R}{J} \ln \frac{v_2}{v_1} = \frac{R}{J} \ln \frac{p_1}{p_2}$$

$$h_2 - h_1 = 0$$

ISENTROPIC (REVERSIBLE ADIABATIC) PROCESSES (CLOSED SYSTEM, IDEAL GAS) (22-8)

$$p_2 = p_1 \left(\frac{v_1}{v_2}\right)^k = p_1 \left(\frac{T_2}{T_1}\right)^{\frac{k}{k-1}}$$

$$T_2 = T_1 \left(\frac{v_1}{v_2}\right)^{k-1} = T_1 \left(\frac{p_2}{p_1}\right)^{\frac{k-1}{k}}$$

$$v_2 = v_1 \left(\frac{p_1}{p_2}\right)^{\frac{1}{k}} = v_1 \left(\frac{T_1}{T_2}\right)^{\frac{1}{k-1}}$$

$$q = 0$$

$$u_2 - u_1 = -W = c_v(T_2 - T_1) = \frac{p_2 v_2 - p_1 v_1}{J(k-1)}$$

$$W = u_1 - u_2 = c_v(T_1 - T_2) = \frac{p_1 v_1 - p_2 v_2}{J(k-1)}$$

$$s_2 - s_1 = 0$$

$$h_2 - h_1 = c_p(T_2 - T_1) = \frac{k(p_2 v_2 - p_1 v_1)}{J(k-1)}$$

ATMOSPHERIC AIR DEFINITIONS (23-1)

The *dry-bulb temperature* is the temperature of still air as measured by a regular thermometer.

The *wet-bulb temperature* is the temperature as measured during an adiabatic-saturation process.

The *dewpoint temperature* is the temperature at which moisture in the air begins to condense out when the air is cooled during a constant-pressure process.

The *humidity ratio (specific humidity)* is the ratio of the mass of water vapor in the air to the mass of the dry air alone.

The *relative humidity* is the ratio of the partial pressure of the vapor to the saturation pressure at the air temperature.

COMPOSITION OF DRY ATMOSPHERIC AIR (30-3)

	% by weight	% by volume
oxygen (O_2)	23.15	20.9
nitrogen (N_2)	76.85	79.1

(rare and inert gases included as N_2)

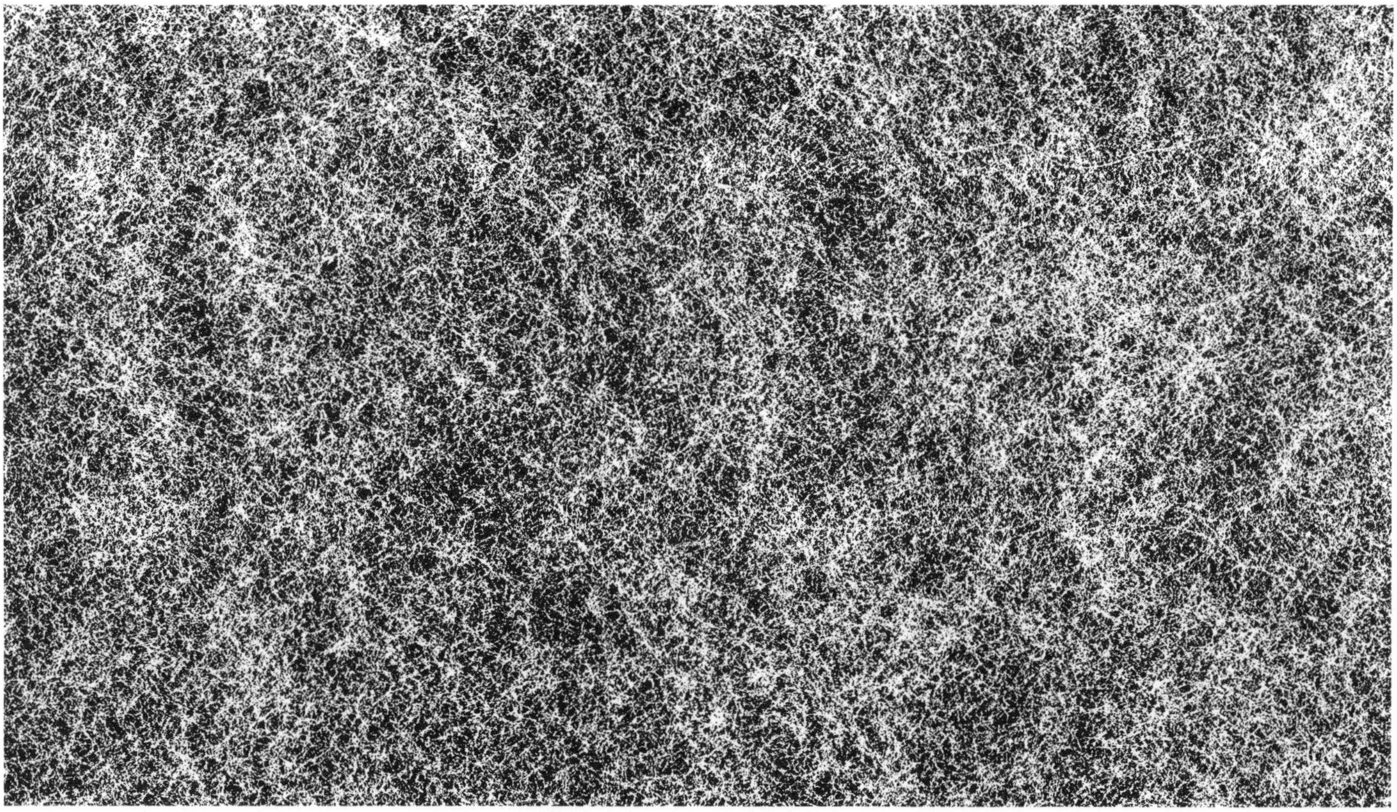

FLOW THROUGH NOZZLES (24-1)

$$v = \sqrt{2g_c J(h_0 - h_2)} \approx 223.8\sqrt{h_0 - h_2}$$

GENERAL POWER CYCLE (25-1)

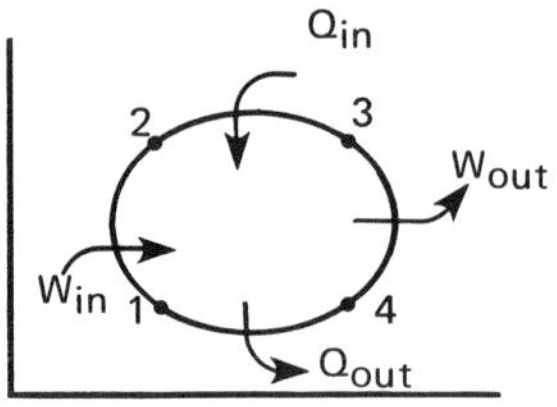

The general power cycle moves *clockwise* on the *p-V* and *T-s* diagrams.

$1 \longrightarrow 2$: compression

$2 \longrightarrow 3$: heat addition

$3 \longrightarrow 4$: expansion

$4 \longrightarrow 1$: heat rejection

THERMAL EFFICIENCY OF THE ENTIRE CYCLE (25-2)

$$\eta_{\text{th}} = \frac{Q_{\text{in}} - Q_{\text{out}}}{Q_{\text{in}}} = \frac{W_{\text{out}} - W_{\text{in}}}{Q_{\text{in}}}$$

ISENTROPIC EFFICIENCY OF A TURBINE (25-5)

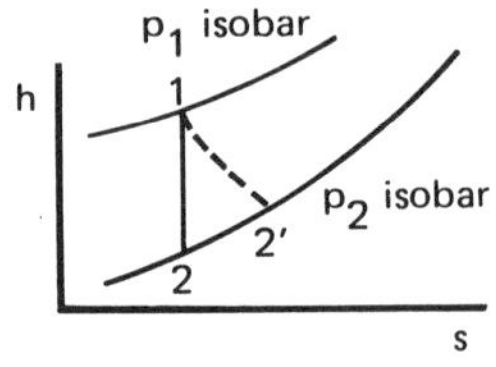

$$\eta_{\text{turbine}} = \frac{h_1 - h_2'}{h_1 - h_2}$$

$$h_2' = h_1 - \eta(h_1 - h_2)$$

ISENTROPIC EFFICIENCY OF A PUMP (25-7)

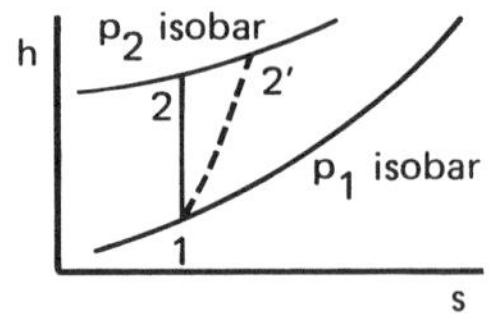

$$\eta_{\text{pump}} = \frac{h_2 - h_1}{h_2' - h_1}$$

$$h_2' = h_1 + \frac{h_2 - h_1}{\eta_{\text{pump}}}$$

CARNOT CYCLE EFFICIENCY (25-8)

For the Carnot cycle, the thermal efficiency does not depend on the working fluid. It can be calculated directly from the two temperature extremes.

$$\eta_{\text{th,Carnot}} = \frac{T_{\text{high}} - T_{\text{low}}}{T_{\text{high}}} \quad [T \text{ in °R or K}]$$

RANKINE CYCLE WITH SUPERHEAT (25-10)

$$W_{\text{turbine}} = h_d - h_e$$

$$W_{\text{pump}} = \frac{v_f(p_a - p_f)}{J} = h_a - h_f$$

$$Q_{\text{in}} = h_d - h_a$$

$$Q_{\text{out}} = h_e - h_f$$

The thermal efficiency of the entire cycle is

$$\eta_{\text{th}} = \frac{Q_{\text{in}} - Q_{\text{out}}}{Q_{\text{in}}} = \frac{W_{\text{turbine}} - W_{\text{pump}}}{Q_{\text{in}}} = \frac{(h_d - h_a) - (h_e - h_f)}{h_d - h_a}$$

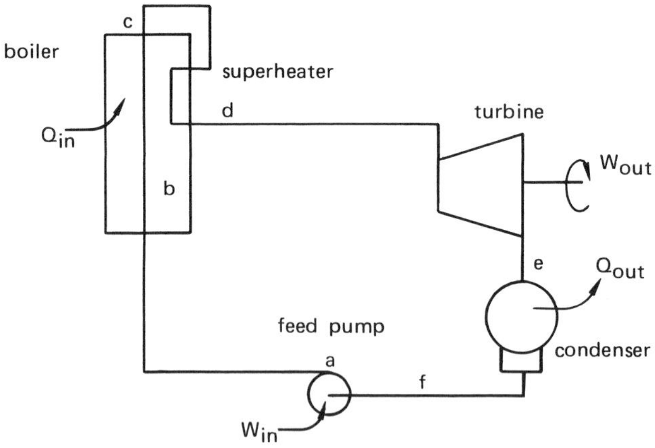

If the efficiencies for the turbine and the pump are known, calculate all quantities as if those efficiencies were 100%. Then, modify h_e and h_a prior to recalculating the thermal efficiency.

$$h_e' = h_d - \eta_{\text{turbine}}(h_d - h_e)$$

$$h_a' = h_f + \frac{h_a - h_f}{\eta_{\text{pump}}}$$

$$W_{\text{turbine}}' = h_d - h_e'$$

$$W_{\text{pump}}' = h_a' - h_f$$

$$Q_{\text{in}}' = h_d - h_a'$$

RATE OF FUEL CONSUMPTION (26-2)

The rate of fuel consumption in internal combustion engines is known as the *specific fuel consumption* (SFC) with units of lbm/hp-hr.

$$\text{hourly fuel consumption} = (\text{hp})(\text{SFC})$$

HORSEPOWER VERSUS TORQUE (26-3)

$$(P_{\text{hp}})(5252) = (T_{\text{ft-lbf}})(n_{\text{rpm}})$$

THE PLAN FORMULA (INTERNAL COMBUSTION ENGINES) (26-5)

$$\text{hp} = \frac{pLAN}{33{,}000}$$

$$N = \frac{(2n)(\text{no. cylinders})}{\text{no. strokes/cycle}}$$

p is the mean effective pressure in psig
L is the stroke length in feet
A is the bore area in in^2
N is the number of power strokes per minute
n is the engine speed in rpm

GENERAL REFRIGERATION CYCLE (27-1)

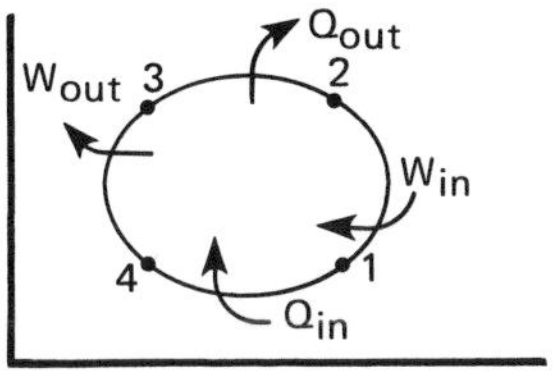

The general refrigeration cycle moves *counterclockwise* on the *p-V* and *T-s* diagrams.

1 → 2 : compression

2 → 3 : heat rejection

3 → 4 : throttling

4 → 1 : heat addition

HEAT PUMPS (27-2)

A heat pump operates on a refrigeration cycle using refrigeration equipment. The only difference is the use to which a heat pump is put. The purpose of a heat pump is to *warm* the region in which the heat rejection coils are located.

COEFFICIENT OF PERFORMANCE (27-2)

Efficiencies are not calculated for refrigeration cycles. Rather, *coefficients of performance* (COP) are used to compare different cycles.

$$\text{COP}_{\text{refrigerator}} = \frac{Q_{\text{absorbed}}}{W_{\text{compression}}} = \frac{Q_{\text{absorbed}}}{Q_{\text{rejected}} - Q_{\text{absorbed}}}$$

$$\text{COP}_{\text{heat pump}} = \frac{Q_{\text{rejected}}}{W_{\text{compression}}} = \frac{Q_{\text{absorbed}} + W_{\text{compression}}}{W_{\text{compression}}} = \text{COP}_{\text{refrigerator}} + 1$$

REFRIGERATION UNITS (27-2)

$$1 \text{ ton} = 200\ \frac{\text{BTU}}{\text{min}} = 12{,}000\ \frac{\text{BTU}}{\text{hr}}$$

CARNOT CYCLE COP (27-2)

$$\text{COP}_{\text{refrigerator}} = \frac{T_{\text{low}}}{T_{\text{high}} - T_{\text{low}}}$$

$$\text{COP}_{\text{heat pump}} = \frac{T_{\text{high}}}{T_{\text{high}} - T_{\text{low}}} = \text{COP}_{\text{refrigerator}} + 1$$

HEAT OF COMBUSTION (30-5)

The heat of combustion of a fuel (also known as the *heating value*) is the amount of energy given off when a unit of fuel is burned. Units of *heating value* are BTU/lbm, BTU/gal, and BTU/ft^3, depending on whether the fuel is solid, liquid, or gas. The *higher heating value* (HHV) includes the heat of vaporization of the water vapor formed.

$$Q_{\text{total}} = m_{\text{fuel}}(\text{HHV}) \quad \text{[water vapor condensed out]}$$

ENERGY, WORK, AND POWER CONVERSIONS (A-1)

multiply	by	to get
BTU	3.929×10^{-4}	hp-hrs
BTU	778.3	ft-lbf
BTU	2.930×10^{-4}	kW-hrs
BTU	1.0×10^{-5}	therms
BTU/hr	0.2161	ft-lbf/sec
BTU/hr	3.929×10^{-4}	hp
BTU/hr	0.2930	W
ft-lbf	1.285×10^{-3}	BTU
ft-lbf	3.766×10^{-7}	kW-hrs
ft-lbf	5.051×10^{-7}	hp-hrs
ft-lbf/sec	4.6272	BTU/hr
ft-lbf/sec	1.818×10^{-3}	hp
ft-lbf/sec	1.356×10^{-3}	kW
hp	2545.0	BTU/hr
hp	550	ft-lbf/sec
hp	0.7457	kW
hp-hr	2545.0	BTU
hp-hr	1.976×10^{6}	ft-lbf
hp-hr	0.7457	kW-hrs
kW	1.341	hp
kW	3412.9	BTU/hr
kW	737.6	ft-lbf/sec

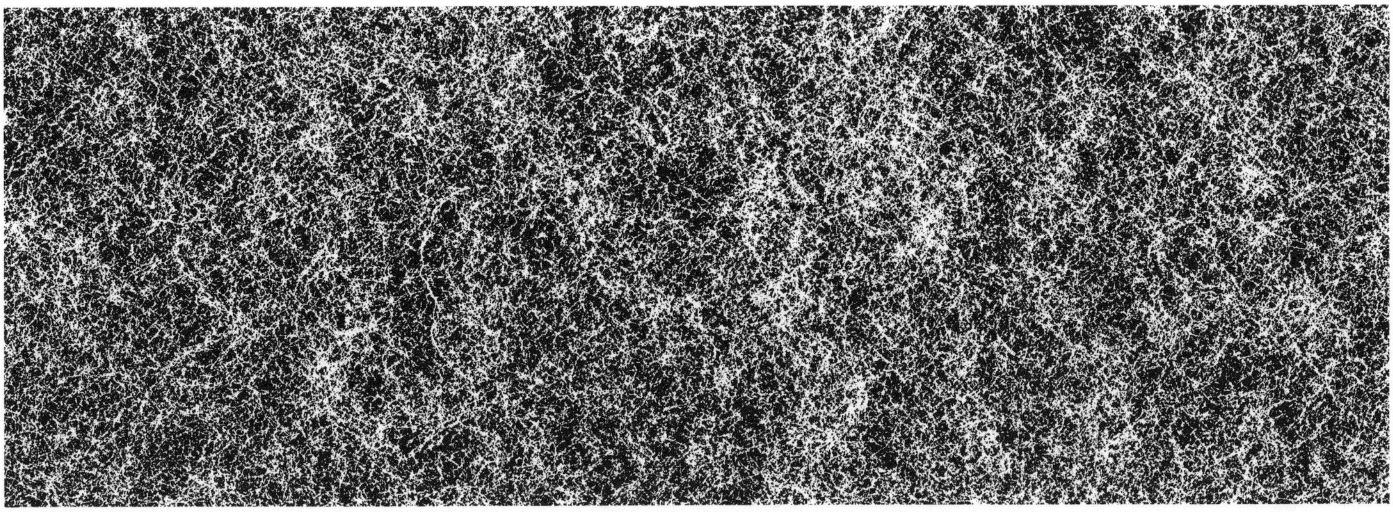

CHEMISTRY

MOLES (29-4)

A mole of a substance is the amount of that substance that has a mass equal to its molecular weight. One gram-mole (mol) of any gas contains 6.022×10^{23} molecules (Avogadro's number) and occupies 22.4 l at scientific STP. One pound-mole (lbmole) of any gas occupies 359.3 ft^3 at scientific STP.

MOLECULAR WEIGHT OF GASES (29-5)

Several important gases form diatomic molecules. These are oxygen (O_2), hydrogen (H_2), chlorine (Cl_2), and nitrogen (N_2).

EQUIVALENT WEIGHT (29-6)

The equivalent weight of an element or radical is its molecular weight divided by its charge when ionized. The equivalent weight in grams supplies one gram-mole (i.e., 6.022×10^{23}) of reacting units.

ELEMENT AND RADICAL TABLE (29-8)

name	symbol	atomic weight	oxidation number
acetate	$C_2H_3O_2$	59.0	−1
aluminum	Al	27.0	+3
ammonium	NH_4	18.0	+1
barium	Ba	137.3	+2
boron	B	10.8	+3
borate	BO_3	58.8	−3
bromine	Br	79.9	−1
calcium	Ca	40.1	+2
carbon	C	12.0	+4, −4
carbonate	CO_3	60	−2
chlorate	ClO_3	83.5	−1
chlorine	Cl	35.5	−1
chlorite	ClO_2	67.5	−1
chromate	CrO_4	116.0	−2
chromium	Cr	52.0	+2, +3, +6
copper	Cu	63.6	+1, +2
dichromate	Cr_2O_7	168	−2
fluorine	F	19.0	−1
gold	Au	197.2	+1, +3
hydrogen	H	1.0	+1
hydroxide	OH	17.0	−1
hypochlorite	ClO	51.5	−1
iron	Fe	55.9	+2, +3
lead	Pb	207.2	+2, +4
lithium	Li	6.9	+1
magnesium	Mg	24.3	+2
mercury	Hg	200.6	+1, +2
nickel	Ni	58.7	+2, +3
nitrate	NO_3	62.0	−1
nitrite	NO_2	46.0	−1
nitrogen	N	14.0	−3, +1, +2, +3, +4, +5
oxygen	O	16	−2
permanganate	MnO_4	118.9	−1
phosphate	PO_4	95.0	−3
phosphorous	P	31.0	−3, +3, +5
potassium	K	39.1	+1
silicon	Si	28.1	+4, −4
silver	Ag	107.9	+1
sodium	Na	23.0	+1
sulfate	SO_4	96.1	−2
sulfite	SO_3	80.1	−2
sulfur	S	32.1	−2, +4, +6
tin	Sn	118.7	+2, +4
zinc	Zn	65.4	+2

OXIDATION-REDUCTION REACTIONS (29-12)

Oxidation occurs when an element or compound loses electrons and becomes less negative (more positive). This occurs at the anode (positive terminal).

Reduction occurs when an element or compound gains electrons and becomes more negative (less positive). This occurs at the cathode (negative terminal).

UNITS OF CONCENTRATION (29-14)

The *normality* (N) is the number of gram-equivalent weights of solute per liter of solution.

The *molarity* (M) is the number of gram-moles of solute per liter of solution.

The *molality* (m) is the number of gram-moles of solute per 1000 g of solvent.

The solution strength may also be expressed in *parts per million* (ppm), which is the number of mass units of solute per million mass units of solution. For water solutions, this is the same as *milligrams per liter* (mg/l).

BOILING POINT/FREEZING POINT CHANGES (29-16)

The boiling point will rise and the freezing point will drop when a solute is added to a solvent. K_b and K_f depend on the molality (m) of solvent only. For water, $K_b = 0.512$ °C/m and $K_f = 1.86$ °C/m.

$$\Delta T_b = mK_b$$

$$\Delta T_f = -mK_f$$

pH AND pOH (29-18)

[X] is the ion concentration of ion X in units of moles per liter.

$$\mathrm{pH} = -\log_{10}[\mathrm{H}^+]$$

$$\mathrm{pOH} = -\log_{10}[\mathrm{OH}^-]$$

$$\mathrm{pH} + \mathrm{pOH} = 14$$

The ion concentration may be calculated from the molarity, M.

$$[\mathrm{X}] = (\text{fraction ionized})(M)$$

ACID/BASE NEUTRALIZATION: TITRATION (29-18)

(acid volume)(acid normality) = (base volume)(base normality)

EQUILIBRIUM CONSTANTS (29-21)

Consider the following reversible reaction.

$$a\mathrm{A} + b\mathrm{B} \rightleftarrows c\mathrm{C} + d\mathrm{D}$$

The equilibrium constant is

$$K_{\mathrm{eq}} = \frac{[\mathrm{C}]^c[\mathrm{D}]^d}{[\mathrm{A}]^a[\mathrm{B}]^b}$$

[X] is the concentration of X.

IONIZATION CONSTANTS (29-21)

$$K_{\mathrm{ion}} = K_{\mathrm{eq}}[\mathrm{H_2O}] = \frac{(\text{molarity})(\text{fraction ionized})^2}{1 - \text{fraction ionized}}$$

ENTHALPY (HEAT) OF REACTION (29-24)

$$\Delta H_r = \sum \Delta H_{f,\text{products}} - \sum \Delta H_{f,\text{reactants}}$$

FARADAY'S LAW OF ELECTROLYSIS (29-25)

The mass of a gas generated by electrolysis is proportional to the amount of electricity used. The name *Faraday* is given to the derived constant which has a value of 96,500 A·s. One Faraday will produce one gram-equivalent weight of a substance through electrolysis.

$$m_{\text{grams}} = \frac{(I_{\text{amps}})(t_{\text{seconds}})(\text{MW})}{(\text{change in oxidation state})(96{,}500)}$$

GALVANIC SERIES (29-29)

(anodic to cathodic, in seawater)

magnesium alloys
zinc
Alclad 3S
aluminum alloys
low-carbon steel
cast iron
stainless – No. 410
stainless – No. 430
stainless – No. 404
stainless – No. 316
Hastelloy A
lead
lead-tin alloys
tin
brass
copper
bronze
90/10 copper-nickel
70/30 copper-nickel
Inconel
silver
stainless steels (passive)
Monel metal
Hastelloy C
titanium
gold

COMPOSITION OF AIR (30-3)

by weight: 23.15% oxygen, 76.85% nitrogen and others
by volume: 20.9% oxygen, 79.1% nitrogen and others

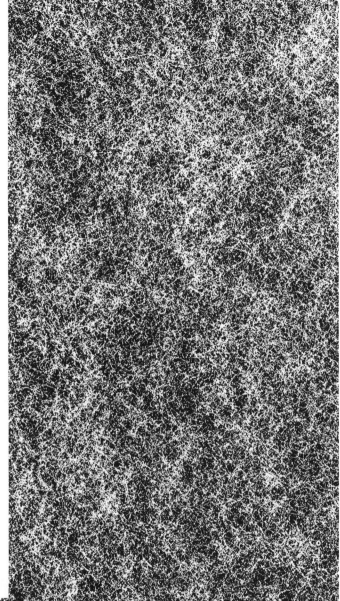

STATICS

RESULTANTS (5-1)

The resultant of n forces with components F_{xi} and F_{yi} has a magnitude of

$$R = \sqrt{(\Sigma F_{xi})^2 + (\Sigma F_{yi})^2}$$

The direction is

$$\phi = \arctan\left(\frac{\Sigma F_{yi}}{\Sigma F_{xi}}\right)$$

FORCES IN VECTOR FORM (5-2)

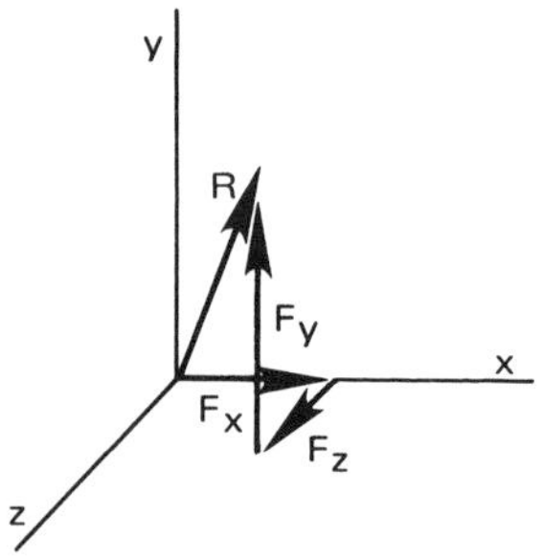

A force, F, can be separated into its components by using the direction cosines.

$$F_x = F\cos\theta_x$$
$$F_y = F\cos\theta_y$$
$$F_z = F\cos\theta_z$$
$$\cos\theta_x = \frac{x}{\sqrt{x^2+y^2+z^2}}$$
$$\cos\theta_y = \frac{y}{\sqrt{x^2+y^2+z^2}}$$
$$\cos\theta_z = \frac{z}{\sqrt{x^2+y^2+z^2}}$$

The resultant can be written in vector form in terms of its components and the unit vectors.

$$\mathbf{F} = \mathbf{i}F_x + \mathbf{j}F_y + \mathbf{k}F_z$$

FORCES (32-2)

A force is a vector quantity. As such, it is completely defined by its magnitude, line of application, sense, and point of application.

COMPONENTS OF FORCES (32-2)

A non-trigonometric method can be used to resolve forces in inclined members into their components. This resolution can be accomplished by multiplying the force by the ratio of sides, as determined from the geometry of the inclined member.

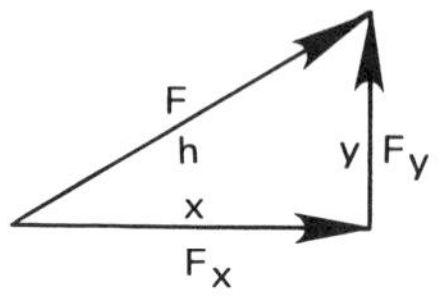

$$F_y = \left(\frac{y}{h}\right)F$$
$$F_x = \left(\frac{x}{h}\right)F$$

MOMENTS (32-2)

The moment produced by a force acting with a moment arm of length d is

$$M = Fd$$

COUPLES (32-4)

A couple is a moment created by two equal forces acting parallel but with opposite directions. The only effect of a couple is to cause rotation. If d is the separation distance between the forces, then

$$M = Fd$$

CONDITIONS FOR EQUILIBRIUM (32-6)

In general, the conditions for equilibrium are

$$\Sigma F_x = 0 \qquad \Sigma F_y = 0 \qquad \Sigma F_z = 0$$
$$\Sigma M_x = 0 \qquad \Sigma M_y = 0 \qquad \Sigma M_z = 0$$

For general coplanar systems, the conditions are

$$\Sigma F_x = 0 \qquad \Sigma F_y = 0 \qquad \Sigma M_z = 0$$

For parallel systems, the conditions are

$$\Sigma F_x = 0 \qquad \Sigma F_y = 0 \qquad \Sigma M_z = 0$$

For concurrent systems, the conditions are

$$\Sigma F_x = 0 \qquad \Sigma F_y = 0$$

SIGN CONVENTIONS FOR TRUSS FORCES (32-12)

For forces in truss members, tension is positive and compression is negative. When freebody diagrams are drawn for truss *joints*, forces leaving the joints place the truss member in tension; forces going into the joints place the member in compression.

DETERMINATE TRUSSES (32-13)

A truss will be determinate if the number of truss members is equal to

$$\text{no. of truss members} = (2)(\text{no. of joints}) - 3$$

If the left-hand side is less than the right-hand side, the truss is not rigid. If the left-hand side is greater than the right-hand side, the truss is indeterminate to the degree of the difference.

PARABOLIC CABLES (32-17)

If the lowest sag point, point B, is used as the origin, the shape of the cable is

$$y(x) = \frac{wx^2}{2H}$$

The approximate length of the cable from the lowest point to the support (i.e., length BD) is

$$L \approx a\left[1 + \tfrac{2}{3}\left(\frac{S}{a}\right)^2 - \tfrac{2}{5}\left(\frac{S}{a}\right)^4\right]$$

The angle of the cable at any point is

$$\tan\theta = \frac{wx}{H}$$

The cable tension is

$$T_{C,x} = H = \frac{wa^2}{2S} \qquad \text{[constant]}$$
$$T_{C,y} = wx$$
$$T_C = \sqrt{(T_{C,x})^2 + (T_{C,y})^2}$$
$$= w\sqrt{\left(\frac{a^2}{2S}\right)^2 + x^2} \qquad \left[\begin{array}{c}\text{minimum at B}\\ \text{maximum at supports}\end{array}\right]$$

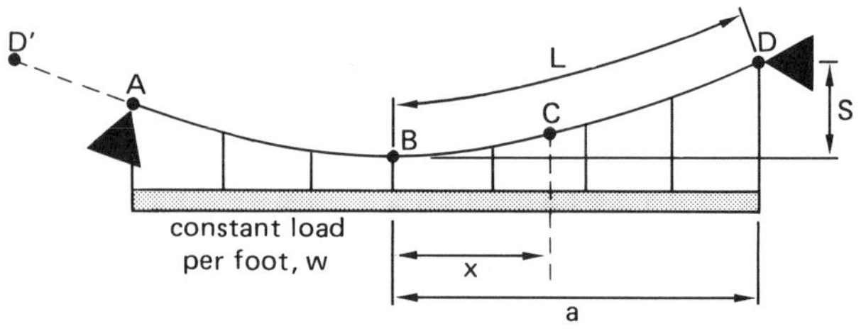

CATENARY CABLES (32-18)

In the following figure and equations, c is the distance from the lowest point on the cable to a reference plane below. The distance c is a constant which must be determined. It does not correspond to any physical dimension.

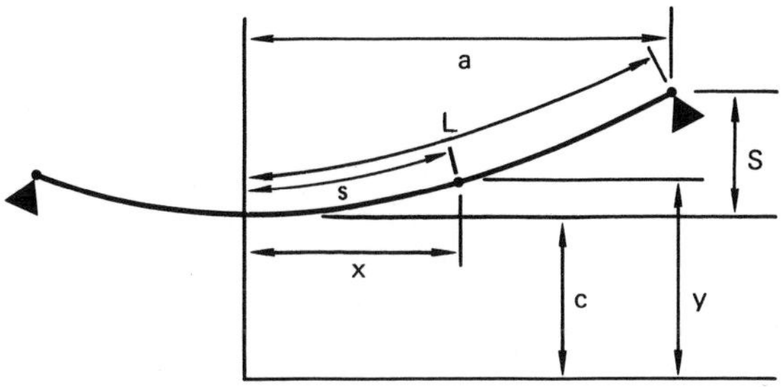

$$y = c\left[\cosh\left(\frac{x}{c}\right)\right]$$

$$s = c\left[\sinh\left(\frac{x}{c}\right)\right]$$

$$y = \sqrt{s^2 + c^2} = c\left[\cosh\left(\frac{x}{c}\right)\right] = c + S$$

$$S = c\left[\cosh\left(\frac{a}{c}\right) - 1\right] = y - c$$

$$\tan\theta = \frac{s}{c}$$

$H = wc$ [horizontal component of tension]

$F = ws$ [applied vertical load due to cable weight]

$T = wy$ [tangential tension]

CENTROIDS (39-1)

The centroid of an object is that point at which the object would balance if suspended. In general, the x and y coordinates of the centroidal location can be found from the following relationships.

$$x_c = \frac{1}{A}\int x\,dA \qquad y_c = \frac{1}{A}\int y\,dA$$

The centroidal locations of common shapes are given below.

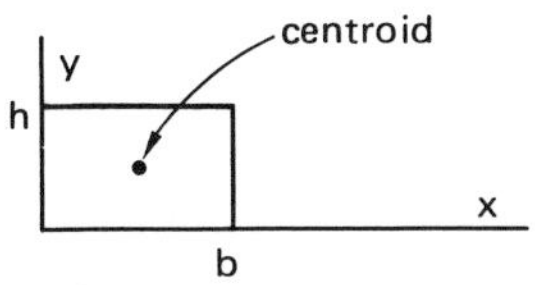

$$x_c = \tfrac{1}{2}b$$
$$y_c = \tfrac{1}{2}h$$

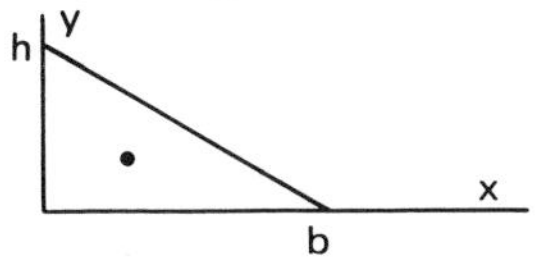

$$x_c = \tfrac{1}{3}b$$
$$y_c = \tfrac{1}{3}h$$

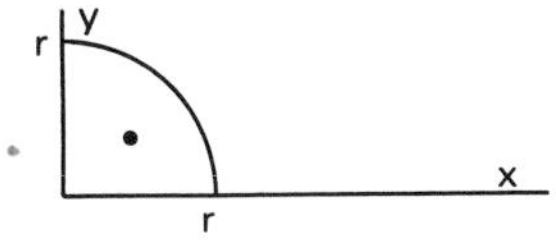

$$x_c = \frac{4r}{3\pi}$$
$$y_c = \frac{4r}{3\pi}$$

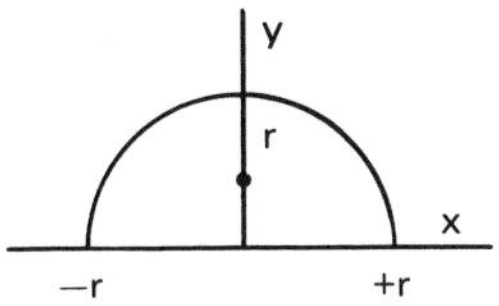

$$x_c = 0$$
$$y_c = \frac{4r}{3\pi}$$

CENTROIDS OF COMPOSITE (BUILT-UP) SHAPES (39-1)

$$x_c = \frac{\Sigma A_i x_{ci}}{\Sigma A_i} \qquad y_c = \frac{\Sigma A_i y_{ci}}{\Sigma A_i}$$

AREA MOMENTS OF INERTIA (39-3)

The moment of inertia (second moment of area) has units of $(\text{length})^4$ and can be considered a measure of resistance to bending. I_x and I_y represent the resistance to bending about the x and y axes, respectively. I_x and I_y are not components.

$$I_x = \int y^2 dA \qquad I_y = \int x^2 dA \qquad dA = dx\,dy$$

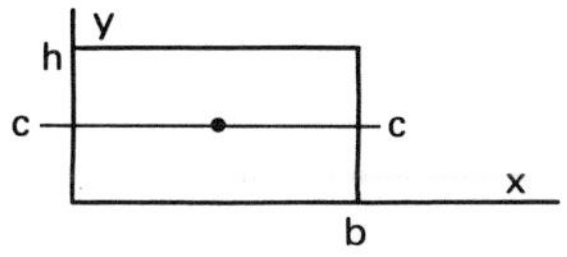

$$I_{x,c} = \tfrac{1}{12}bh^3$$
$$I_x = \tfrac{1}{3}bh^3$$

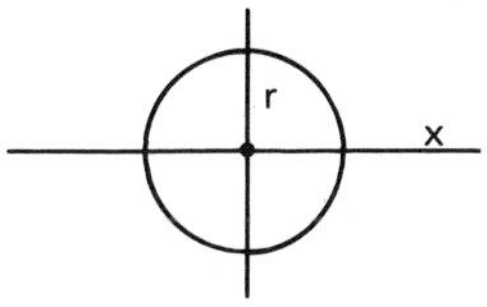

$$I_{x,c} = \tfrac{1}{4}\pi r^4$$

PARALLEL AXIS THEOREM (39-4)

If the moment of inertia about the centroidal axis is known, the moment of inertia about a second parallel axis located at distance d away from the centroidal axis is

$$I_{\text{new}} = I_{\text{centroidal}} + Ad^2 \quad \text{[area]}$$

POLAR MOMENTS OF INERTIA (39-5)

In general, the polar moment of inertia is

$$J = \int r^2 dA = \int (x^2 + y^2)\,dA = I_x + I_y$$

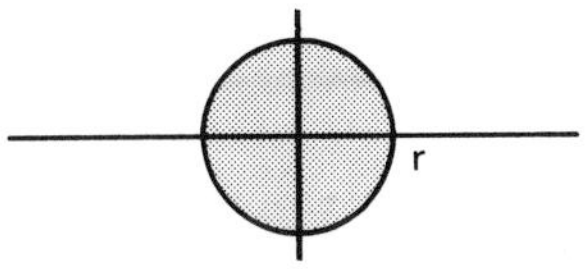

$$J = \tfrac{1}{2}\pi r^4 = \tfrac{1}{32}\pi d^4$$

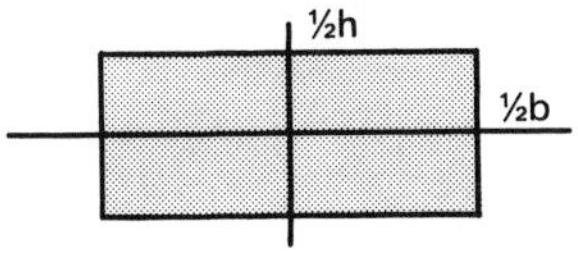

$$J = \tfrac{1}{12}bh(b^2 + h^2)$$

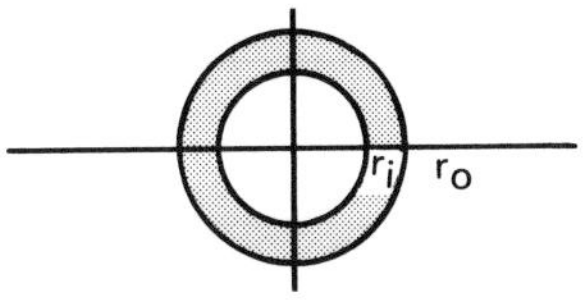

$$J = \tfrac{1}{2}\pi(r_o^4 - r_i^4)$$

RADIUS OF GYRATION (39-6)

The radius of gyration is the distance from a reference axis at which all of the area can be considered to be concentrated to produce the actual moment of inertia.

$$k = \sqrt{\frac{I}{A}} \qquad r = \sqrt{\frac{J}{A}} \quad \text{[for polar moments of inertia]}$$

FRICTION (44-5)

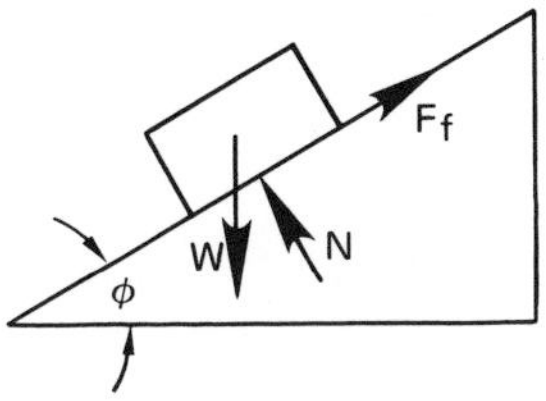

The frictional force depends on the coefficient of friction (f) and the normal force (N).

$$F_f = fN$$

$$N = \frac{mg\cos\phi}{g_c}$$

$$\text{frictional work} = F_f \times \text{distance}$$

STEEL DESIGNATIONS (34-5)

(XX is the carbon content, 0.XX%)

alloy number	major alloying elements
10XX	plain carbon steel
11XX	resulfurized plain carbon
13XX	manganese
23XX, 25XX	nickel
31XX, 33XX	nickel, chromium
40XX	molybdenum
41XX	chromium, molybdenum
43XX	nickel, chromium, molybdenum
46XX, 48XX	nickel, molybdenum
51XX	chromium
61XX	chromium, vanadium
81XX, 86XX, 87XX	nickel, chromium, molybdenum
92XX	silicon

ALUMINUM DESIGNATIONS (34-10)

alloy number	major alloying ingredient
1XXX	commercially pure aluminum (99%+)
2XXX	copper
3XXX	manganese
4XXX	silicon
5XXX	magnesium
6XXX	magnesium and silicon
7XXX	zinc
8XXX	other

ALUMINUM TEMPERS (34-10)

temper	description
T2	annealed (castings only)
T3	solution heat-treated, followed by cold working
T4	solution heat-treated, followed by natural aging
T5	artificial aging
T6	solution heat-treated, followed by artificial aging
T7	solution heat-treated, followed by stabilizing by overaging heat treating
T8	solution heat-treated, followed by cold working and subsequent artificial aging

CRYSTALLINE STRUCTURE (35-2)

Crystalline structures are categorized into 14 different point-lattices. Important structures are simple cubic, body-centered cubic (BCC), face-centered cubic (FCC), simple hexagonal, and hexagonal close-packed (HCP).

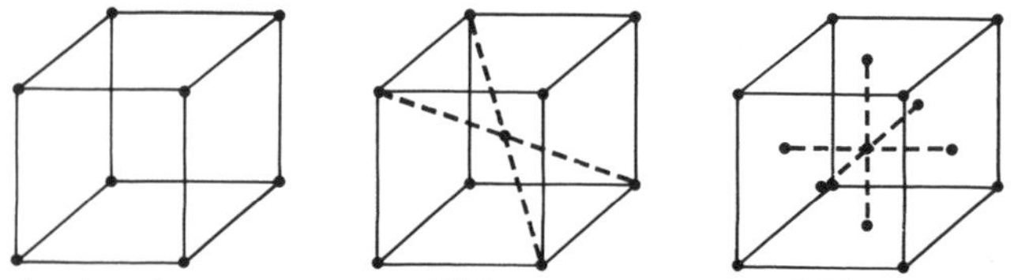

simple cube BCC FCC

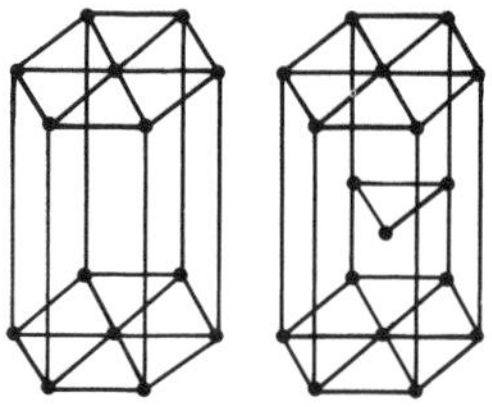

simple hexagonal HCP

Common Metals (35-2)

The following metals are BCC: chromium, alpha iron, lithium, molybdenum, potassium, sodium, tantalum, beta titanium, and alpha tungsten.

The following metals are FCC: aluminum, alpha brass, copper, gold, gamma iron, lead, nickel, platinum, and silver.

The following metals are HCP: beryllium, cadmium, magnesium, alpha titanium, and zinc.

Cell Packing Parameters (35-4)

(assuming hard touching spheres of radius r)

type of cell	number of atoms	packing factor	coordination number
simple cubic	1	0.52	6
BCC	2	0.68	8
FCC	4	0.74	12
simple hexagonal	2	0.52	8
HCP	6	0.74	12

NUMBER OF ATOMS IN A CELL (35-3)

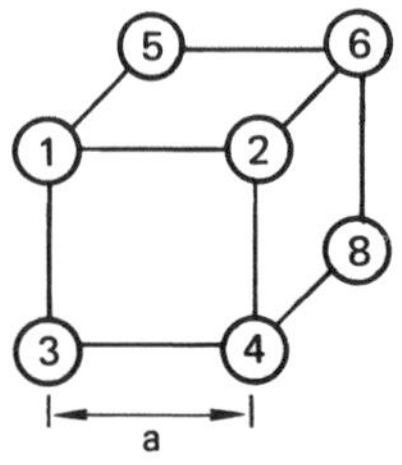

distance between atoms	in terms of r	in terms of a
1 and 2	$2r$	a
1 and 4	$2\sqrt{2}r$	$\sqrt{2}a$
1 and 8	$2\sqrt{3}r$	$\sqrt{3}a$

(a) simple cubic

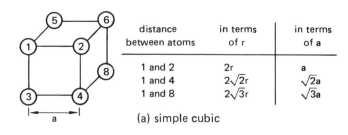

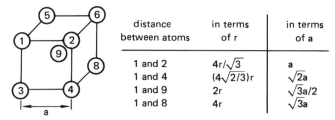

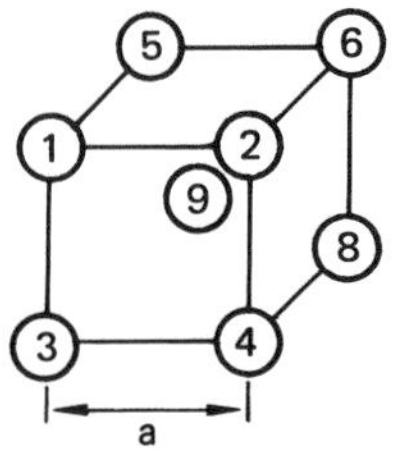

distance between atoms	in terms of r	in terms of a
1 and 2	$4r/\sqrt{3}$	a
1 and 4	$(4\sqrt{2/3})r$	$\sqrt{2}a$
1 and 9	$2r$	$\sqrt{3}a/2$
1 and 8	$4r$	$\sqrt{3}a$

(b) body-centered cubic

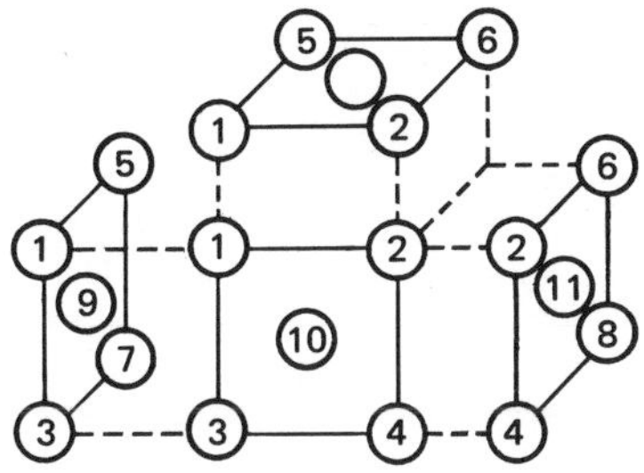

distance between atoms	in terms of r	in terms of a
1 and 2	$2\sqrt{2}r$	a
1 and 10	2r	$\sqrt{2}a/2$
1 and 4	4r	$\sqrt{2}a$
1 and 8	$2\sqrt{6}r$	$\sqrt{3}a$
1 and 11	$2\sqrt{3}r$	$\sqrt{3/2}a$
10 and 11	2r	$\sqrt{2}a/2$
9 and 11	$2\sqrt{2}r$	a

(c) face-centered cubic

CRYSTALLINE DIRECTIONS (35-5)

Crystalline directions are specified by a system based essentially on a coordinate system defined by the unit cell. A direction which starts from (0,0,0) and intersects the unit cell at (u, v, w) is specified as $[uvw]$. No commas are used. Negative numbers are written as positive numbers with overbars.

CRYSTALLINE PLANES: MILLER INDICES (35-5)

A plane satisfies the equation $\frac{x}{a}+\frac{y}{b}+\frac{z}{c}=1$ where a, b, and c are the intercepts of the plane on the x, y, and z axes, respectively. If a, b, and c are the intercepts on the axes within a unit cell, the *Miller indices* are written as (hkl) where $h=\frac{1}{a}$, $k=\frac{1}{b}$, and $l=\frac{1}{c}$. No commas are used. Negative numbers are written as positive numbers with overbars.

X-RAY DIFFRACTION (35-6)

If x-rays of wavelength λ are used to find the spacing (d_{hkl}) of crystalline planes, the relationship between the diffraction angle and the nth reinforcement is

$$n\lambda = (2)(d_{hkl})\sin\theta \quad \text{[Bragg's law]}$$

$$\text{Angstrom units:}\quad \text{Å} = 10^{-8}\ \text{cm} = 10^{-10}\ \text{m}$$

SPACING BETWEEN CRYSTALLINE PLANES (35-6)

Planar spacing in cubic systems is

$$d_{hkl} = \frac{\mathbf{a}}{\sqrt{h^2+k^2+l^2}}$$

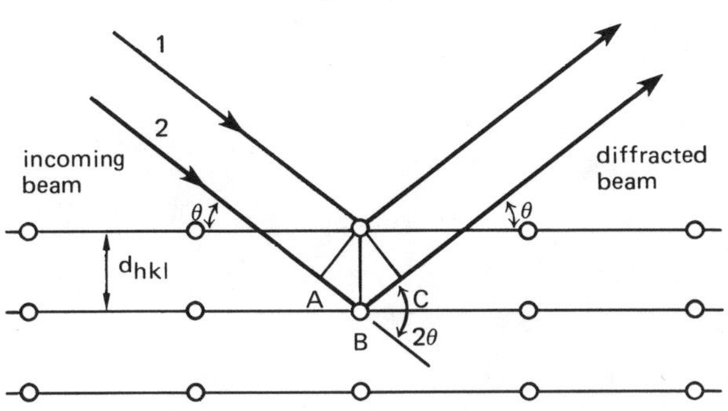

POINT DEFECTS (35-7)

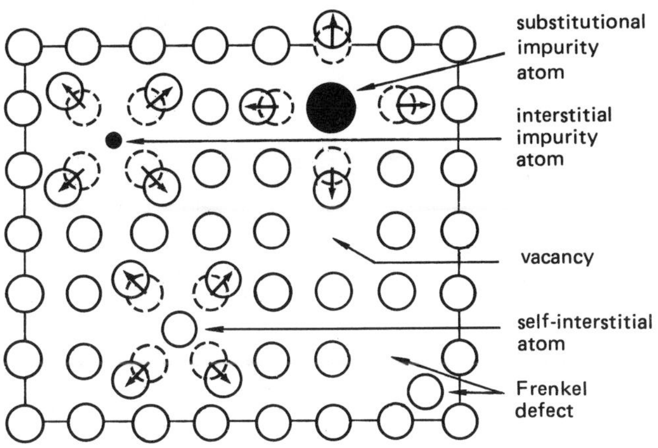

FICK'S LAWS OF DIFFUSION (35-8)

$$J = -D\frac{dC}{dx}$$

$$\frac{dC}{dt} = D\frac{d^2C}{dx^2}$$

TENSILE TEST (36-1)

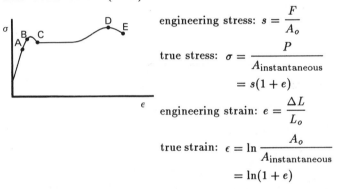

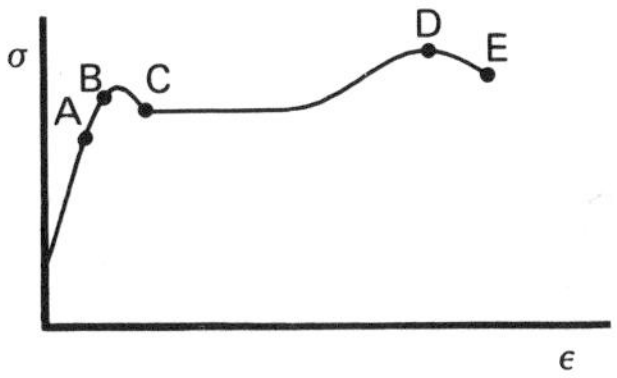

engineering stress: $s = \dfrac{F}{A_o}$

true stress: $\sigma = \dfrac{P}{A_{\text{instantaneous}}} = s(1+e)$

engineering strain: $e = \dfrac{\Delta L}{L_o}$

true strain: $\epsilon = \ln\dfrac{A_o}{A_{\text{instantaneous}}} = \ln(1+e)$

point A: proportionality limit—the highest stress for which Hooke's law ($\sigma = E\epsilon$) is valid.

point B: elastic limit—the highest stress for which no permanent deformation occurs.

point C: yield point—the stress at which a sharp drop in load-carrying ability occurs.

point D: ultimate strength—the highest stress which the material can achieve.

point E: fracture strength—the stress at fracture

The *modulus of elasticity (E)* is the slope of the line for stresses up to the proportionality limit.

The *shear modulus (G)* can be calculated from the modulus of elasticity and Poisson's ratio.

$$G = \frac{E}{2(1+\nu)}$$

The *modulus of toughness* is the work per unit volume required to cause fracture. It is calculated as the area under the σ-ϵ curve up to the point of fracture. Units of toughness are in-lbf/in^3.

The *ductility* is a measure of the amount of plastic strain at the breaking point. It is calculated as the percent reduction in area at fracture.

$$\text{reduction in area} = \frac{A_o - A_{\text{fracture}}}{A_o} = \frac{L_{\text{fracture}} - L_o}{L_o}$$

The *percent elongation at fracture* is calculated from the original length and the fracture length after the sample has "snapped back."

FATIGUE TEST (36-8)

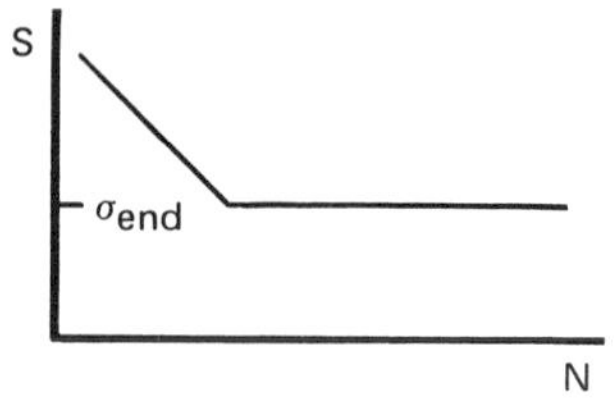

Endurance tests (fatigue tests) apply a cyclical loading of constant maximum amplitude. The plot (usually semi-log or log-log) of the maximum stress and the number of cycles to failure is known as an *S-N curve.*

The *endurance stress (endurance limit* or *fatigue limit)* is the maximum stress which can be repeated indefinitely without causing failure.

The *fatigue life* is the number of cycles required to cause failure for a given stress level.

HARDNESS VERSUS ULTIMATE STRENGTH (36-9)

For heat-treated plain-carbon and medium-alloy steels,

$$S_{ut} \approx (500)(\text{BHN}) \quad \text{[in psi]}$$

IMPACT (TOUGHNESS) TEST (36-11)

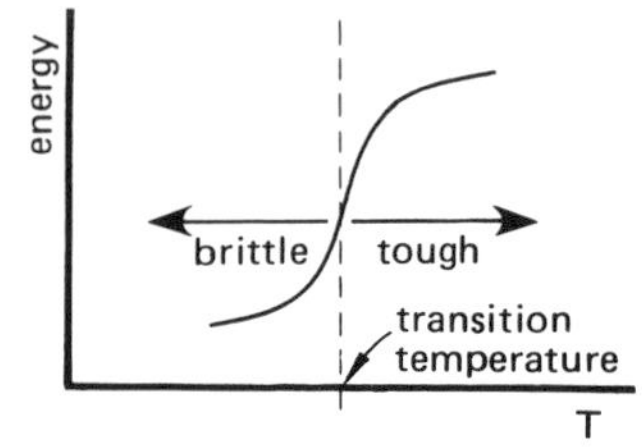

Impact tests determine the amount of energy required to cause failure in standardized test samples. The tests are repeated over a range of temperatures to determine the *transition temperature.* (The transition temperature is approximately 30°F for low-carbon steel.)

CREEP TEST (36-12)

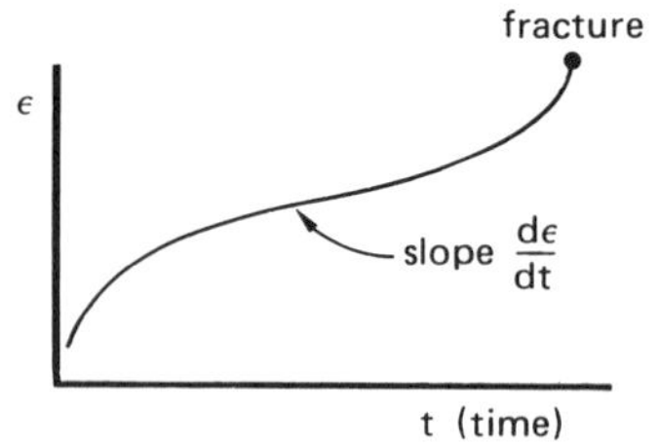

A constant stress less than the yield strength is applied and the elongation versus time is measured.

The *creep rate* ($d\epsilon/dt$) is very temperature dependent.

The *creep strength* is the stress which results in a given creep rate.

The *rupture strength* is the stress which results in failure after some given amount of time.

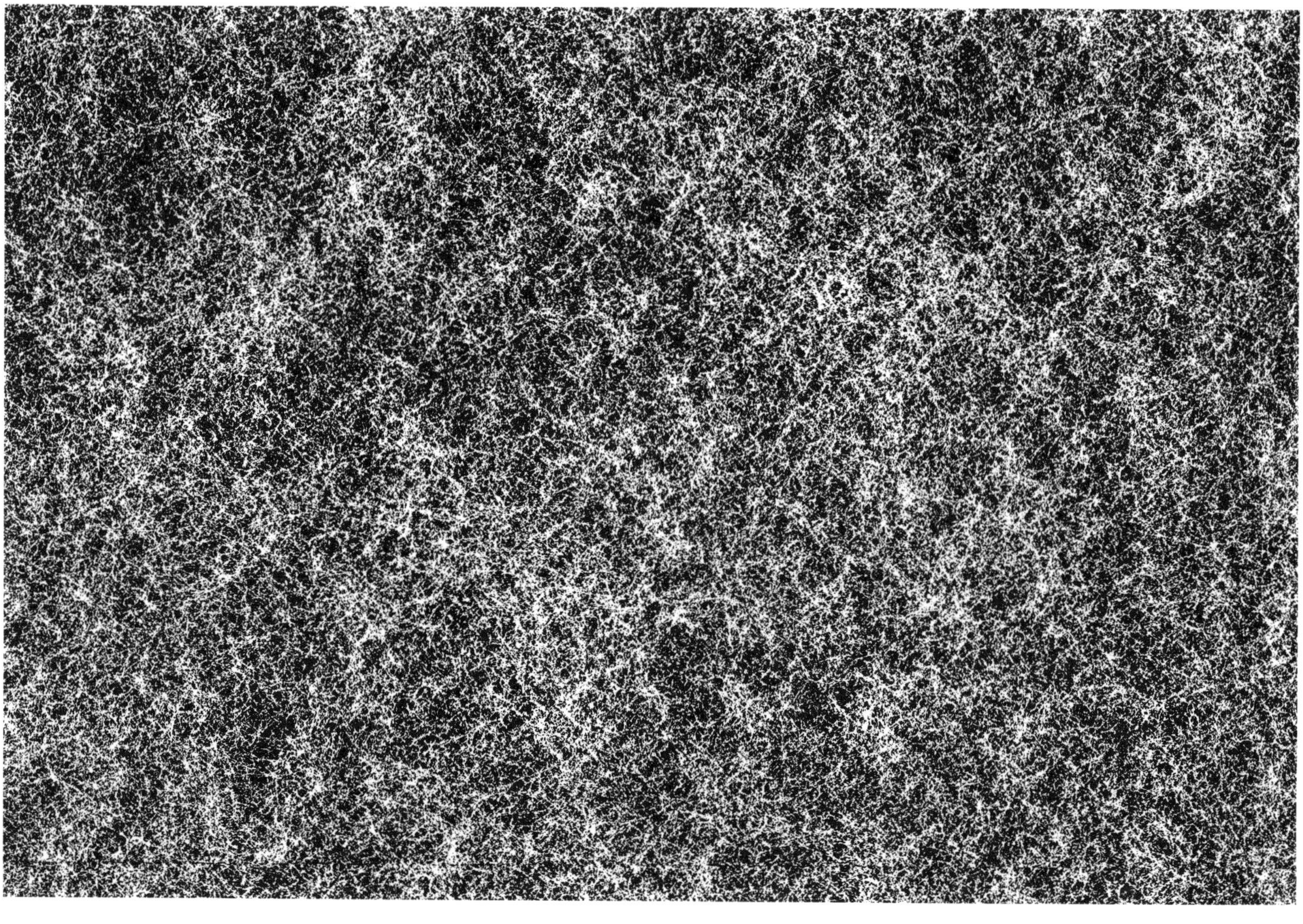

PROPERTIES OF STEEL

modulus of elasticity (E):	2.9×10^7 psi 2×10^5 MPa
modulus of shear (G):	1.2×10^7 psi 7.9×10^4 MPa
Poisson's ratio (ν):	0.3
coefficient of thermal expansion (α):	6.5×10^{-6} 1/°F 1.2×10^{-6} 1/°C
density (ρ):	489 lbm/ft^3 7830 kg/m^3

PROPERTIES OF ALUMINUM

modulus of elasticity (E):	1×10^7 psi 7×10^4 MPa
modulus of shear (G):	3.9×10^7 psi 2.6×10^4 MPa
Poisson's ratio (ν):	0.33
coefficient of thermal expansion (α):	1.3×10^{-5} 1/°F 2×10^{-5} 1/°C
density (ρ):	173 lbm/ft^3 2770 kg/m^3

POISSON'S RATIO (36-4)

ν is the ratio of traverse strain to longitudinal strain. (F in y direction)

$$\nu = \frac{\Delta D L_o}{D_o \Delta L} = \frac{\epsilon_x}{\epsilon_y}$$

RELATIONSHIP BETWEEN SHEAR AND ELASTIC MODULI (36-8)

$$G = \frac{E}{2(1+\nu)}$$

STRESS (40-2)

Normal stress: $\sigma = E\epsilon = \dfrac{F}{A}$

Shear stress: $\tau = G\theta = \dfrac{F}{A}$

STRAIN (F in y direction) (40-2)

$$\epsilon_y = \frac{\Delta L}{L_o} = \frac{F}{AE}$$

$$\epsilon_x = \nu \epsilon_y$$

$$\sigma_y = E\epsilon_y$$

HOOKE'S LAW (40-2)

$\sigma = E\epsilon$ [normal stress]

$\tau = G\theta$ [shear stress]

ELONGATION UNDER NORMAL STRESS (40-2)

$$\Delta L = L_o \epsilon = \frac{F L_o}{AE}$$

THERMAL STRESS AND STRAIN (40-3)

The elongation of an unconstrained object when heated is

$$\Delta L = \alpha L_o \Delta T$$

The thermal stress and strain are

$$\epsilon_{\text{th}} = \alpha \Delta T$$

$$\sigma_{\text{th}} = E\epsilon_{\text{th}} = \alpha E \Delta T$$

COMBINED STRESS (40-4)

The normal and shear stresses on a plane whose normal is inclined an angle θ from the horizontal are

$$\sigma_\theta = \tfrac{1}{2}(\sigma_x + \sigma_y) + \tfrac{1}{2}(\sigma_x - \sigma_y)\cos 2\theta + \tau \sin 2\theta$$

$$\tau_\theta = -\tfrac{1}{2}(\sigma_x - \sigma_y)\sin 2\theta + \tau \cos 2\theta$$

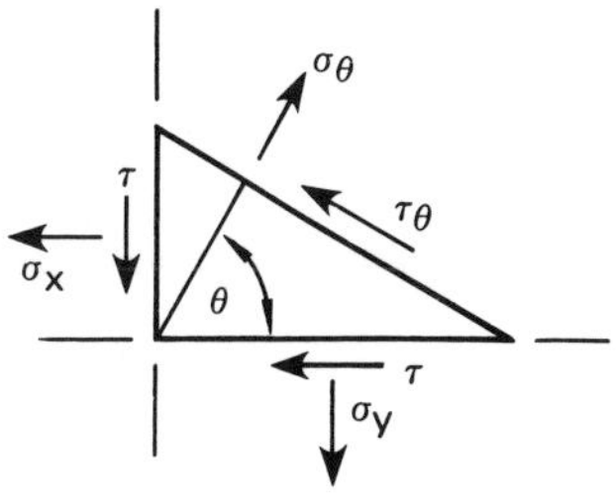

The maximum and minimum values of σ_θ and τ_θ (as θ is varied) are the *principal stresses*. These are

$$\sigma_{\text{max,min}} = \tfrac{1}{2}(\sigma_x + \sigma_y) \pm \tau_{\text{max}}$$

$$\tau_{\text{max,min}} = \pm\tfrac{1}{2}\sqrt{(\sigma_x - \sigma_y)^2 + (2\tau)^2}$$

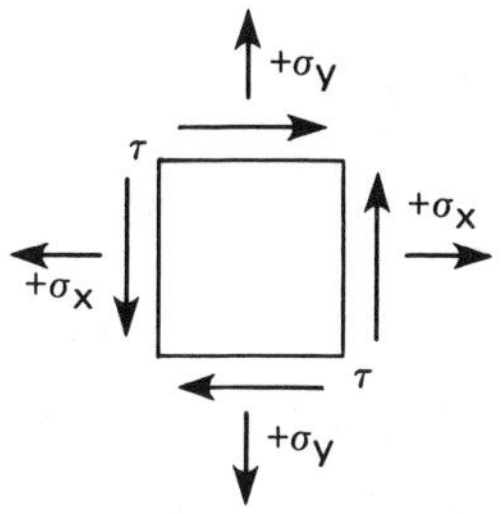

Proper sign convention must be adhered to. Normal tensile stresses are positive; normal compressive stresses are negative. Shear stresses are positive as shown.

PRINCIPAL STRESSES (40-5)

$$\theta_{\sigma_1,\sigma_2} = \tfrac{1}{2}\arctan\left(\frac{2\tau}{\sigma_x - \sigma_y}\right)$$

$$\theta_{\tau_1,\tau_2} = \tfrac{1}{2}\arctan\left(\frac{\sigma_x - \sigma_y}{-2\tau}\right)$$

SHEAR DIAGRAMS (40-8)

- Loads and reactions acting upward are positive.
- Shear at any point is the sum of forces up to that point.
- Concentrated loads produce horizontal straight lines.
- Uniform loads produce straight inclined lines.
- Shear at any point is the slope of the moment diagram at that point. That is, $V = dM/dx$.

MECHANICS OF MATERIALS

MOMENT DIAGRAMS (40-8)

- Clockwise moments are positive. (Use the left-hand rule.)
- Concentrated loads produce straight inclined lines.
- Uniform loads produce parabolic lines.
- Maximum moment occurs where shear (V) is zero.
- Moment is zero at a free end or hinge.
- Moment at any point is the area under the shear diagram up to that point. That is, $M = \int V dx$.

SHEAR STRESS IN BEAMS (40-9)

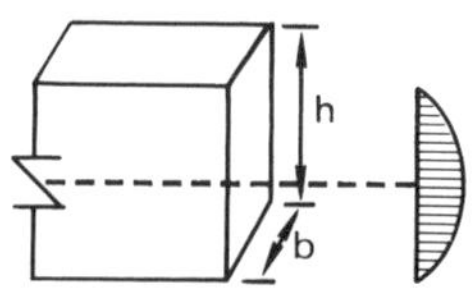

$$\tau = \frac{VQ}{Ib}$$

$$\tau_{max} = \frac{3V}{2A} = \frac{3V}{2bh} \quad \text{[rectangle]}$$

$$\tau_{max} = \frac{4V}{3A} = \frac{4V}{3\pi r^2} \quad \text{[circle]}$$

V = shear

Q = statical moment

BENDING STRESS IN BEAMS (40-10)

Normal stress in a beam is also called "flexure stress" and "bending stress."

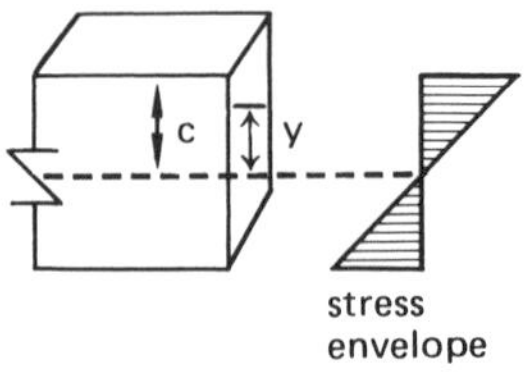

$$\sigma = \frac{My}{I_c}$$

$$\sigma_{max} = \frac{Mc}{I_c} = \frac{M}{Z}$$

$$I_c = \frac{bh^3}{12} \quad \text{[rectangle]}$$

$$I_c = \frac{\pi d^4}{64} \quad \text{[circle]}$$

$$Z = \text{section modulus} = \frac{I_c}{c}$$

ECCENTRIC NORMAL STRESS (40-11)

$$\sigma_{max}, \sigma_{min} = \frac{F}{A} \pm \frac{Mc}{I_c} = \frac{F}{A} \pm \frac{Fec}{I_c}$$

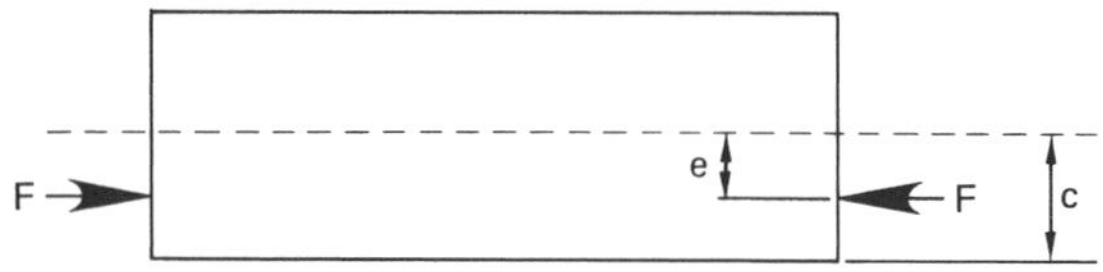

BEAM SLOPE/DEFLECTION RELATIONS (40-13)

y = deflection

$$y' = \frac{dy}{dx} = \text{slope}$$

$$y'' = \frac{d^2y}{dx^2} = \frac{M(x)}{EI}$$

$$y''' = \frac{d^3y}{dx^3} = \frac{V(x)}{EI}$$

SIMPLE BEAM DEFLECTIONS (A-80) (App. 40.A)

(w = load/unit of length)

type of beam	loading	M_{max}	deflection
cantilever	at tip	FL	$\frac{FL^3}{3EI}$
cantilever	uniform	$\frac{1}{2}wL^2$	$\frac{wL^4}{8EI}$
simple	at center	$\frac{1}{4}FL$	$\frac{FL^3}{48EI}$
simple	uniform	$\frac{wL^2}{8}$	$\frac{5wL^4}{384EI}$

ALLOWABLE STRESS (41-2)

The allowable stress may be calculated from either the yield stress (typical for ductile materials like steel) or the ultimate strength (typical for brittle materials like cast iron).

$$\sigma_{allowable} = \frac{S_{yield}}{\text{factor of safety}} \quad \text{[ductile]}$$

$$= \frac{S_{ultimate}}{\text{factor of safety}} \quad \text{[brittle]}$$

SLENDER COLUMNS IN COMPRESSION (41-3)

Euler's formula may be used if the actual stress is kept less than the yield strength. Slender columns fail by buckling.

$$\text{slenderness ratio} = \frac{\text{unbraced length}}{\text{radius of gyration}}$$

$$F_e = \frac{\pi^2 EI}{(CL)^2}$$

$$F_{max} = \frac{F_e}{\text{factor of safety}}$$

$$\sigma_e = \frac{F_e}{A}$$

Theoretical End Restraint Coefficients

end conditions	C
both ends pinned	1
both ends built in	0.5
one end pinned, one end built in	0.707
one end built in, one end free	2
one end built in, one end fixed against rotation but free	1
one end pinned, one end fixed against rotation but free	2

SPRINGS (41-4)

The force required to elongate a spring with a spring constant of k is

$$F = kx \quad \text{[Hooke's law]}$$

$$W = \tfrac{1}{2}kx^2$$

$$k = k_1 + k_2 + k_3 + \cdots \quad \text{[parallel springs]}$$

$$\frac{1}{k} = \frac{1}{k_1} + \frac{1}{k_2} + \frac{1}{k_3} + \cdots \quad \text{[series springs]}$$

THIN-WALLED TANKS (PIPE STRESS) (41-4)

A tank has thin walls if the wall thickness is less than 1/10 of the tank diameter.

$$\sigma_{\text{hoop}} = \frac{pr}{t}$$

r = tank radius

t = wall thickness

p = gage pressure inside tank

$$\sigma_{\text{longitudinal}} = \frac{pr}{2t}$$

For spherical tanks or spherical ends on a cylindrical tank, use the longitudinal stress formula.

TORSIONAL STRESS AND STRAIN IN A SHAFT (41-7)

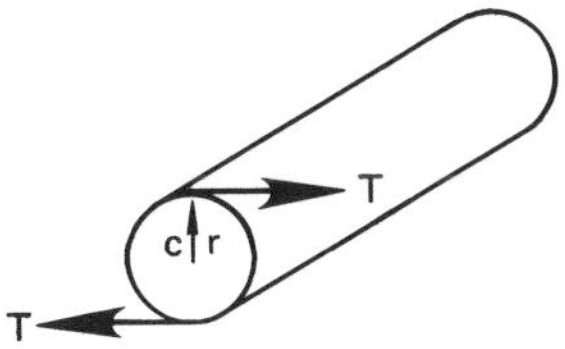

$$\tau = \frac{Tr}{J}$$

$$\tau_{\text{max}} = \frac{Tc}{J}$$

J = polar (area) moment of inertia

$= \frac{1}{2}\pi r^4$ [round shafts]

$$\phi = \frac{TL}{GJ} \quad \text{[in radians]}$$

The maximum torque that the shaft can carry is

$$T_{\text{max}} = \frac{\tau_{\text{max}} J}{r}$$

The horsepower transmitted by the shaft is

$$\text{hp} = \frac{T_{\text{in-lbf}}\, n_{\text{rpm}}}{63{,}025}$$

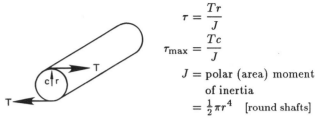

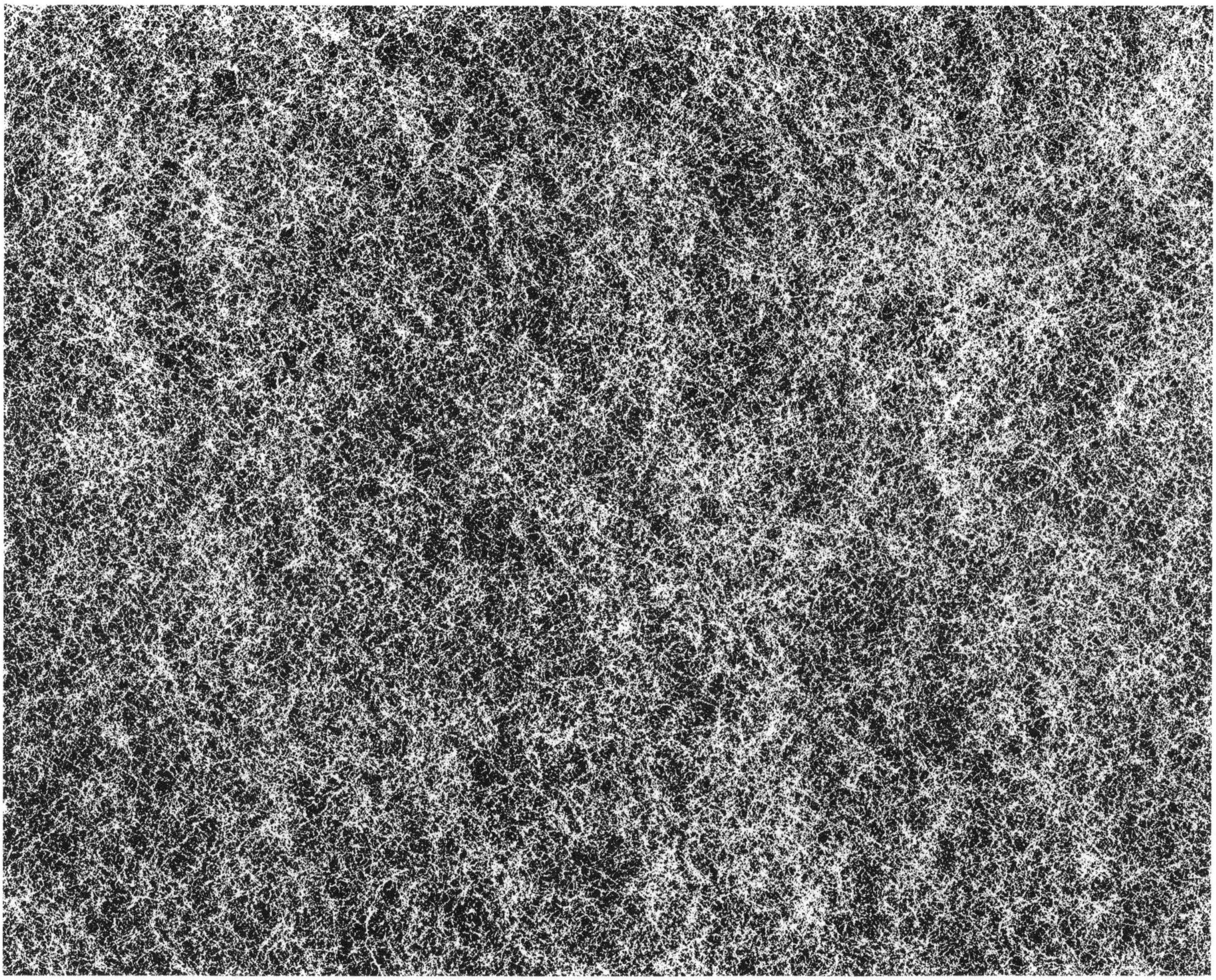

DYNAMICS

CONVERSIONS (A-1)

1 mph = 1.467 ft/sec

WORK AND ENERGY (20-2)

Work is an energy transfer that occurs when a force, F, (or torque, T) is moved through a distance (an angle).

$$\mathbf{W} = \int \mathbf{F} \cdot d\mathbf{s} \quad \text{[linear]}$$

$$\mathbf{W} = \int \mathbf{T} \cdot d\theta \quad \text{[rotational]}$$

Potential energy is the energy that an object possesses by virtue of its position in a gravitational field.

$$E_p = \frac{mgh}{g_c}$$

Kinetic energy is the energy that an object possesses by virtue of its velocity.

$$E_k = \frac{m\text{v}^2}{2g_c} \quad \text{[linear]}$$

$$E_k = \frac{I\omega^2}{2g_c} \quad \text{[rotational]}$$

For a spring (with ***stiffness*** k) compressed an amount x,

$$E_p = \tfrac{1}{2}kx^2$$

THE WORK-ENERGY PRINCIPLE (20-4)

In the absence of thermal changes, the work done on (or by) a system is equal to its change in energy.

$$W = \Delta E_p + \Delta E_k$$

POWER (20-5)

$$P = \frac{W}{\Delta t} = F\text{v} = T\omega$$

HORSEPOWER REQUIRED TO MAINTAIN VELOCITY (20-5)

For translation, $\text{hp} = \dfrac{F\text{v}_{\text{ft/sec}}}{550}$

For rotation, $\text{hp} = \dfrac{2\pi T_{\text{ft-lbf}} n_{\text{rpm}}}{33{,}000}$

SYSTEMS OF SPRINGS (41-4)

For springs in parallel,

$$k_{\text{equivalent}} = k_1 + k_2 + k_3 + \cdots$$

For springs in series,

$$\frac{1}{k_{\text{equivalent}}} = \frac{1}{k_1} + \frac{1}{k_2} + \frac{1}{k_3} + \cdots$$

For two springs in series,

$$k_{\text{equivalent}} = \frac{k_1 k_2}{k_1 + k_2}$$

MASS MOMENT OF INERTIA FOR ROTATIONAL SYSTEMS (42-2)

In general, $I = \int r^2 dm$

For a solid cylinder, $I = \dfrac{mr^2}{2g_c}$

For a hollow cylinder, $I = \dfrac{m(r_i^2 + r_o^2)}{2g_c}$

PARALLEL AXIS THEOREM (42-2)

$$I_{\text{new}} = I_{\text{centroidal}} + \frac{md^2}{g_c}$$

RADIUS OF GYRATION (42-2)

$$k = \sqrt{\frac{Ig_c}{m}}$$

DISTANCE, VELOCITY, AND ACCELERATION (43-2)

$$a = \frac{d\text{v}}{dt} = \frac{d^2s}{dt^2}$$

$$\text{v} = \frac{ds}{dt} = \int a\,dt$$

$$s = \int \text{v}\,dt = \int\int a\,dt^2$$

UNIFORM ACCELERATION (43-3)

For translation,

$$\text{v} = \text{v}_0 + at$$

$$s = \text{v}_0 t + \frac{at^2}{2}$$

$$\text{v}^2 = \text{v}_0^2 + 2as$$

For rotation,

$$\omega = \omega_0 + \alpha t$$

$$\theta = \omega_0 t + \frac{\alpha t^2}{2}$$

$$\omega^2 = \omega_0^2 + 2\alpha\theta$$

UNIFORM ACCELERATION FORMULAS (43-4)

to find	given these	use this formula
t	a, v_0, v	$t = \dfrac{\text{v} - \text{v}_0}{a}$
t	a, v_0, s	$t = \dfrac{\sqrt{2as + \text{v}_0^2} - \text{v}_0}{a}$
t	v_0, v, s	$t = \dfrac{2s}{\text{v}_0 + \text{v}}$
a	t, v_0, v	$a = \dfrac{\text{v} - \text{v}_0}{t}$
a	t, v_0, s	$a = \dfrac{2s - 2\text{v}_0 t}{t^2}$
a	v_0, v, s	$a = \dfrac{\text{v}^2 - \text{v}_0^2}{2s}$
v_0	t, a, v	$\text{v}_0 = \text{v} - at$
v_0	t, a, s	$\text{v}_0 = \dfrac{s}{t} - \tfrac{1}{2}at$
v_0	a, v, s	$\text{v}_0 = \sqrt{\text{v}^2 - 2as}$
v	t, a, v_0	$\text{v} = \text{v}_0 + at$
v	a, v_0, s	$\text{v} = \sqrt{\text{v}_0^2 + 2as}$
s	t, a, v_0	$s = \text{v}_0 t + \tfrac{1}{2}at^2$
s	a, v_0, v	$s = \dfrac{\text{v}^2 - \text{v}_0^2}{2a}$
s	t, v_0, v	$s = \tfrac{1}{2}t(\text{v}_0 + \text{v})$

PROJECTILE MOTION (43-4)

Falling body problems may be solved with the above uniform acceleration formulas by substituting

$$a = g = 32.2 \text{ ft/sec}^2 \; (9.8 \text{ m/s}^2)$$

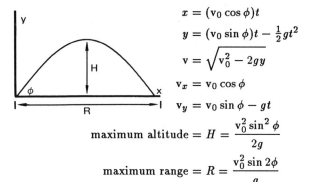

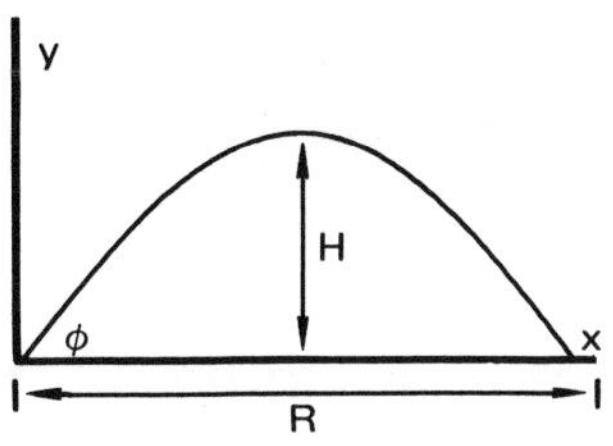

$$x = (\mathrm{v}_0 \cos\phi)t$$

$$y = (\mathrm{v}_0 \sin\phi)t - \tfrac{1}{2}gt^2$$

$$\mathrm{v} = \sqrt{\mathrm{v}_0^2 - 2gy}$$

$$\mathrm{v}_x = \mathrm{v}_0 \cos\phi$$

$$\mathrm{v}_y = \mathrm{v}_0 \sin\phi - gt$$

$$\text{maximum altitude} = H = \frac{\mathrm{v}_0^2 \sin^2\phi}{2g}$$

$$\text{maximum range} = R = \frac{\mathrm{v}_0^2 \sin 2\phi}{g}$$

$$\text{total flight time} = T = \frac{2\mathrm{v}_0 \sin\phi}{g}$$

The above formulas neglect air drag. Range, R, is maximum when $\phi = 45°$.

ROTATIONAL MOTION (43-6)

α : rotational acceleration
ω : rotational speed
θ : rotational position

$$\alpha = \frac{d\omega}{dt} = \frac{d^2\theta}{dt^2}$$

$$\omega = \frac{d\theta}{dt} = \int \alpha\, dt$$

$$\theta = \int \omega\, dt = \int\int \alpha\, dt^2$$

INSTANTANEOUS RELATIONSHIP BETWEEN ROTATIONAL AND LINEAR MOTION (43-7)

$$\omega = 2\pi\left(\frac{\text{rpm}}{60}\right)$$

$$s = r\theta$$

$$\mathrm{v} = r\omega$$

$$a = r\alpha$$

NEWTON'S LAWS (44-3, 44-5, 44-18)

First law: The velocity (momentum) of an object will not change unless it is acted upon by a force.

Second law: $\mathbf{F} = \frac{d\mathbf{p}}{dt} = \frac{md\mathbf{v}}{dt} = m\mathbf{a}$ [consistent units]

Third law: For every action there is an equal but opposite reaction.

Law of Universal Gravitation: The gravitational force between two objects with masses m_1 and m_2 is

$$F = \frac{Gm_1m_2}{r^2} \quad \text{[consistent units]}$$

NEWTON'S SECOND LAW FOR ROTATIONAL SYSTEMS (44-4)

$$\mathbf{T} = \frac{d\mathbf{h}}{dt} = I\frac{d\omega}{dt} = I\alpha$$

CENTRIFUGAL FORCE (44-4)

$$F_c = \frac{m\mathrm{v}^2}{g_c r}$$

$$a_n = \frac{\mathrm{v}_t^2}{r} = r\omega^2 = \mathrm{v}_t\omega$$

ROADWAY BANKING (44-7)

The banking angle (superelevation) needed on a circular curve of radius r is

$$\phi = \arctan\frac{\mathrm{v}^2}{gr}$$

IMPULSE-MOMENTUM PRINCIPLE (44-14)

For translational motion, $F\Delta t = \dfrac{m\Delta \mathrm{v}}{g_c}$

For rotational motion, $T\Delta t = \dfrac{I\Delta\omega}{g_c}$

[I is mass moment of inertia]

COLLISIONS (44-15)

The *coefficient of restitution*, e, is 1 if the collision is perfectly elastic. e is 0 if the collision is completely inelastic (i.e., the bodies stick together).

$$e = \frac{\mathrm{v}_1' - \mathrm{v}_2'}{\mathrm{v}_2 - \mathrm{v}_1}$$

Momentum is conserved in all collisions, although kinetic energy may not be ($e < 1$). The conservation of momentum equation is

$$m_1\mathrm{v}_1' + m_2\mathrm{v}_2' = m_1\mathrm{v}_1 + m_2\mathrm{v}_2$$

SIMPLE HARMONIC MOTION (45-2)

$$\omega = 2\pi f = \frac{2\pi}{T}$$

$$f = \frac{\omega}{2\pi} = \frac{1}{T}$$

$$T = \frac{1}{f} = \frac{2\pi}{\omega}$$

For a simple spring-and-mass oscillator,

$$f = \frac{1}{2\pi}\sqrt{\frac{kg_c}{m}}$$

$$T = 2\pi\sqrt{\frac{m}{kg_c}}$$

For a simple pendulum consisting of a mass, m, at the end of a massless cord of length L,

$$f = \frac{1}{2\pi}\sqrt{\frac{g}{L}}$$

$$T = 2\pi\sqrt{\frac{L}{g}}$$

DC ELECTRICITY

ENERGY CONVERSIONS (A-1)

1 J = 1 N·m = 1 W·s = 10^7 ergs

POWER CONVERSIONS (A-1)

1 W = 1 J/s
1 hp = 745.7 W

PREFIXES (1-7)

pico (p) = 10^{-12} micro (μ) = 10^{-6} kilo (k) = 10^3
nano (n) = 10^{-9} milli (m) = 10^{-3} meg (M) = 10^6

CHARGE (46-2)

The charge on an electron is 1 electrostatic unit (esu), equal to 1.602×10^{-19} C. 1 C is equal to 6.24×10^{18} esu.

ELECTROSTATICS (46-2 to 46-7)

flux: ψ (in lines) $\quad \psi = Q$

flux density: $$\sigma = \frac{\psi}{A} = \frac{Q}{A}$$

For a point charge, $\sigma = \dfrac{Q}{4\pi r^2}$

electric field: $$E = \frac{D}{\epsilon}$$

$\epsilon = \epsilon_0 \epsilon_r$
= permittivity of the medium

$\epsilon_0 = 8.854 \times 10^{-12}\ \mathrm{C^2/N{\cdot}m^2}$ [vacuum]

ϵ_r = dielectric constant (1 for vacuum)

For a point charge, $E = \dfrac{Q}{4\pi\epsilon r^2}$

Coulomb's law: $$F = Q_2 E = \frac{Q_1 Q_2}{4\pi\epsilon r^2}$$

work: The work required to change the separation of two charges from r_1 to r_2 is

$$W = \frac{-Q_1 Q_2}{4\pi\epsilon}\left(\frac{1}{r_2} - \frac{1}{r_1}\right)$$

system potential energy: $$U_p = \frac{-Q_1 Q_2}{4\pi\epsilon r}$$

potential: $$V = \frac{U_p}{Q}$$

UNIFORM ELECTRIC FIELDS (46-4)

If a voltage V appears across two plates separated by a distance r, then the electric field is $E = V/r$. $V = Er$

The force that a charged particle feels is $F = EQ$.

The work done in moving a particle a distance d in the uniform field is $W = Fd = EQd = VQd/r$.

MAGNETISM (46-8)

flux: ϕ (in Wb) $= BA = m$

flux density: $$B = \frac{\phi}{A}$$

For a pole of strength m, $B = \dfrac{m}{4\pi r^2}$

magnetic field: $$H = \frac{B}{\mu}$$

$\mu = \mu_0 \mu_r$
= permeability of the medium

$\mu_0 = 4\pi \times 10^{-7}$ H/m

force: $$F = m_2 H = \frac{m_1 m_2}{4\pi\mu r^2}$$

MAGNETIC FIELD AROUND A WIRE CARRYING CURRENT (46-9)

$$H = \frac{B}{\mu} = \frac{I}{2\pi r} \quad [r \text{ is distance from wire center}]$$

$$F = IBl$$

FARADAY'S LAW (46-10)

The induced voltage in n coils is

$$V = -N\frac{d\phi}{dt} = \frac{d\lambda}{dt} \quad [\lambda \text{ is the } \textit{flux linkage}]$$

RESISTIVITY (47-1)

The resistance of a material whose resistivity is ρ and whose length and area are l and A, respectively, is

$$R = \frac{\rho l}{A}$$

A can be measured in circular mils. A circular mil is the area of a 0.001-in diameter circle.

$$A_{\text{cmils}} = \left(\frac{d_{\text{inches}}}{0.001}\right)^2$$

Resistance varies with temperature according to

$$R = R_0(1 + \alpha\Delta T)$$

$$\rho = \rho_0(1 + \alpha\Delta T)$$

RESISTORS IN SERIES AND PARALLEL (47-2)

series: $$R_{\text{equivalent}} = R_1 + R_2 + R_3 + \cdots$$

parallel: $$\frac{1}{R_{\text{equivalent}}} = \frac{1}{R_1} + \frac{1}{R_2} + \frac{1}{R_3} + \cdots$$

For 2 resistors in parallel: $R_{\text{equivalent}} = \dfrac{R_1 R_2}{R_1 + R_2}$

OHM'S LAW (47-3)

$$V = IR$$

POWER IN A RESISTIVE CIRCUIT (47-4)

$$P = IV = I^2 R = \frac{V^2}{R}$$

DECIBELS (47-5)

$$\text{dB} = 10 \log_{10}\frac{P_2}{P_1}$$

$$20 \log_{10}\frac{V_2}{V_1} \quad [\text{if impedances are the same}]$$

DC ELECTRICITY

KIRCHHOFF'S LAWS (47-5)

Kirchhoff's current law: $\Sigma I_{\text{in}} = \Sigma I_{\text{out}}$ [at a junction]
Kirchhoff's voltage law: $\Sigma V = \Sigma(IR)$ [around a loop]

DELTA-WYE TRANSFORMATIONS (47-6)

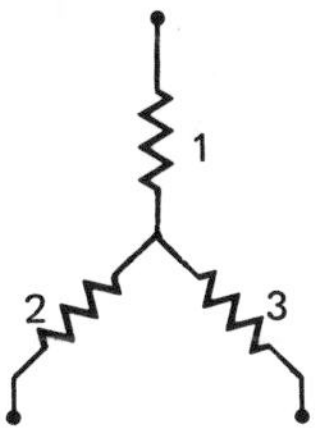

$$R = R_a + R_b + R_c$$
$$R_1 = \frac{R_a R_c}{R}$$
$$R_2 = \frac{R_a R_b}{R}$$
$$R_3 = \frac{R_b R_c}{R}$$

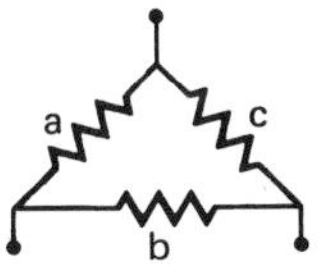

$$R = R_1R_2 + R_1R_3 + R_2R_3$$
$$R_a = \frac{R}{R_3}$$
$$R_b = \frac{R}{R_1}$$
$$R_c = \frac{R}{R_2}$$

NORTON EQUIVALENT (47-7)

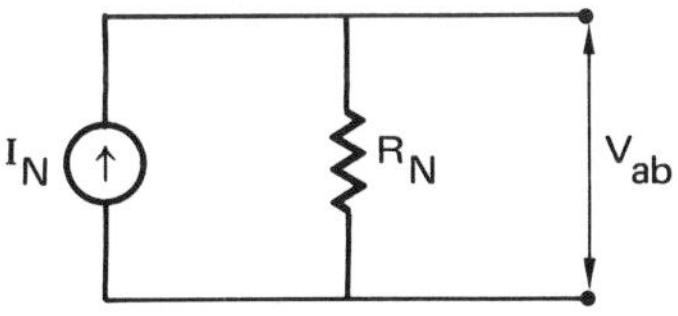

$$V_{ab} = I_N R_N$$

THEVENIN EQUIVALENT (47-8)

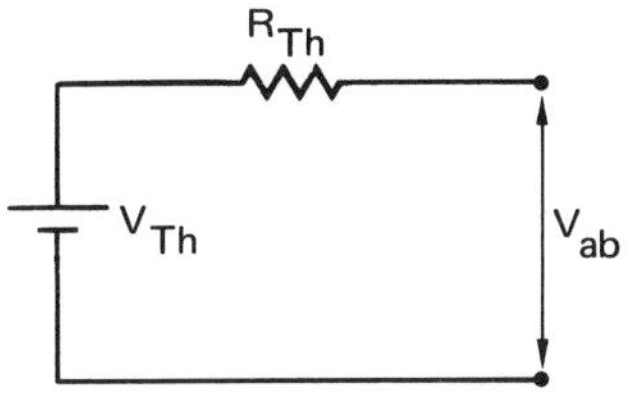

$$V_{ab} = I_{\text{Th}} R_{\text{Th}}$$

WHEATSTONE BRIDGE (47-10)

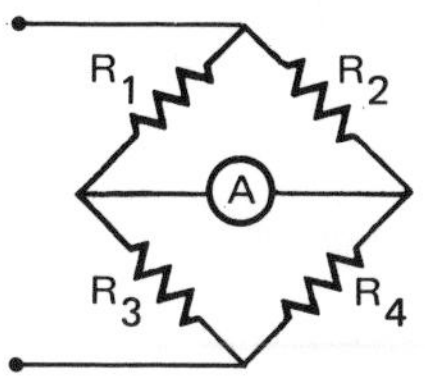

When the ammeter reads zero,

$$I_2 = I_4$$
$$I_1 = I_3$$
$$V_1 + V_3 = V_2 + V_4$$
$$\frac{R_1}{R_2} = \frac{R_3}{R_4}$$

DC TRANSIENT SOLUTIONS (47-10)

type of circuit	response
series *RC*, charging $\tau = RC$ $e^{-N} = e^{-t/\tau} = e^{-t/RC}$ 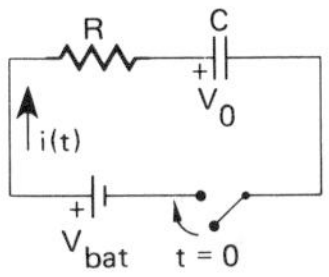	$V_{\text{bat}} = v_R(t) + v_C(t)$ $i(t) = \frac{V_{\text{bat}} - V_0}{R} e^{-N}$ $v_R(t) = i(t)R = (V_{\text{bat}} - V_0)e^{-N}$ $v_C(t) = V_0 + (V_{\text{bat}} - V_0)(1 - e^{-N})$ $Q_C(t) = C[V_0 + (V_{\text{bat}} - V_0) \times (1 - e^{-N})]$
series *RC*, discharging $\tau = RC$ $e^{-N} = e^{-t/\tau} = e^{-t/RC}$ 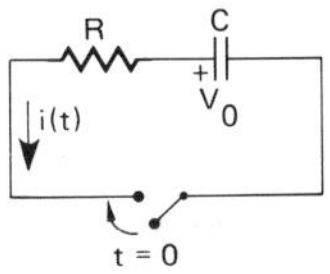	$0 = v_R(t) + v_C(t)$ $i(t) = \frac{V_0}{R} e^{-N}$ $v_R(t) = -V_0 e^{-N}$ $v_C(t) = V_0 e^{-N}$ $Q_C(t) = CV_0 e^{-N}$
series *RL*, charging $\tau = L/R$ $e^{-N} = e^{-t/\tau} = e^{-tR/L}$ 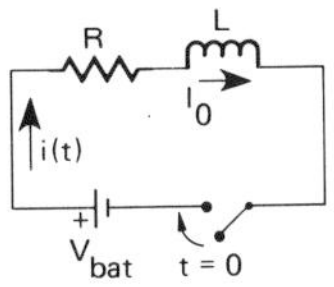	$V_{\text{bat}} = v_R(t) + v_C(t)$ $i(t) = I_0 e^{-N} + \frac{V_{\text{bat}}}{R}(1 - e^{-N})$ $v_R(t) = i(t)R$ $= I_0 R e^{-N} + V_{\text{bat}}(1 - e^{-N})$ $v_L(t) = (V_{\text{bat}} - I_0 R)e^{-N}$
series *RL*, discharging $\tau = L/R$ $e^{-N} = e^{-t/\tau} = e^{-tR/L}$ 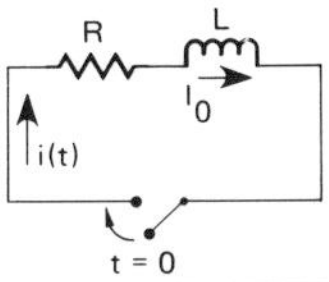	$0 = v_R(t) + v_L(t)$ $i(t) = I_0 e^{-N}$ $v_R(t) = I_0 R e^{-N}$ $v_L(t) = -I_0 R e^{-N}$

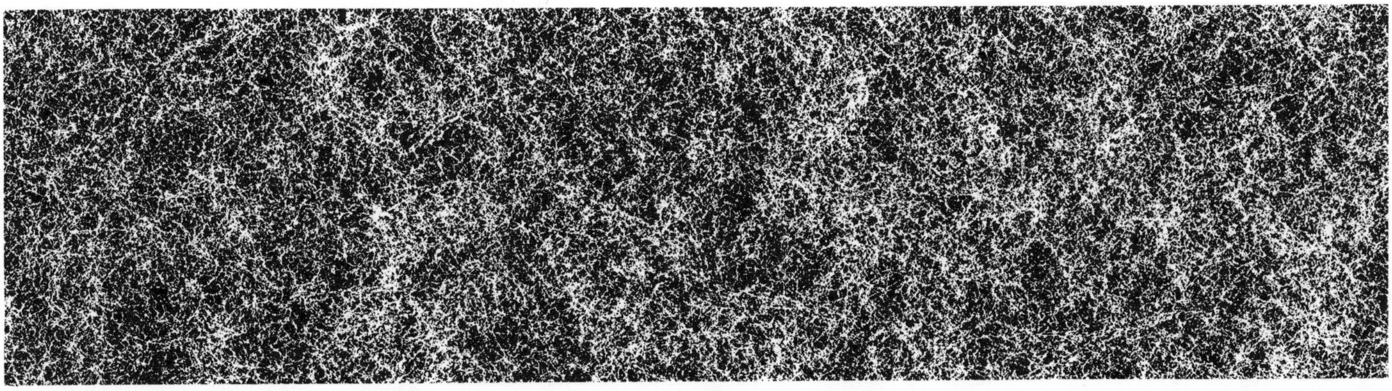

AC ELECTRICITY

SINE-COSINE CONVERSIONS (6-3)

$$\cos(\omega t) = \sin(\omega t + \tfrac{\pi}{2}) = -\sin(\omega t - \tfrac{\pi}{2})$$
$$\sin(\omega t) = \cos(\omega t - \tfrac{\pi}{2}) = -\cos(\omega t + \tfrac{\pi}{2})$$

CAPACITORS (ICE – LEADING) (46-2)

governing equations: $Q(t) = CV(t)$

$$V(t) = \left(\frac{1}{C}\right)\int I(t)\,dt$$

$$I = \frac{dQ(t)}{dt}$$

capacitance: $C = \dfrac{\epsilon A}{r} = \dfrac{\epsilon_r \epsilon_0 A}{r}$

ϵ_0 = 8.854×10^{-12} F/m (same as $C^2/N{\cdot}m^2$)
A = plate area in m^2
r = plate separation in m

reactance: $X_C = \dfrac{1}{\omega C} = \dfrac{1}{2\pi f C}$

angle: $\angle -90°$

power: $P(t) = \frac{1}{2} I_m V_m \sin 2\omega t$

average power: $P_{ave} = 0$

capacitors in series: $\dfrac{1}{C} = \sum \dfrac{1}{C_i}$

$$V_{C_1} = \frac{C_2}{C_1 + C_2} \times V_{total}$$

capacitors in parallel: $C = \sum C_i$

INDUCTORS (ELI – LAGGING) (46-11)

governing equation: $V(t) = L\dfrac{dI(t)}{dt}$

reactance: $X_L = \omega L = 2\pi f L$

angle: $\angle 90°$

power: $P(t) = -\frac{1}{2} I_m V_m \cos 2\omega t$

average power: $P_{ave} = 0$

inductors in series: $L = \sum L_i$

inductors in parallel: $\dfrac{1}{L} = \sum \dfrac{1}{L_i}$

TIME, PERIOD, AND FREQUENCY (48-2)

$$\omega = 2\pi f = \frac{2\pi}{T}$$
$$f = \frac{\omega}{2\pi} = \frac{1}{T}$$
$$T = \frac{2\pi}{\omega} = \frac{1}{f}$$

AVERAGE VALUES (48-3)

For a repeating waveform with period T, the *average value* is

$$V_{ave} = \frac{1}{T}\int_0^T V(t)\,dt$$

The average value of a rectified sinusoid is

$$V_{ave} = \frac{2V_m}{\pi} \quad \text{[rectified sinusoid]}$$

EFFECTIVE VALUES (48-3)

For a repeating waveform with period T, the *effective value* (also known as *root-mean-square* value) is

$$V_{eff} = \sqrt{\frac{1}{T}\int_0^T V^2(t)\,dt}$$

The effective value of a sinusoid is

$$\frac{V_m}{\sqrt{2}} = 0.707 V_m$$

FORM AND CREST FACTOR (48-4)

The *form factor* is V_{eff}/V_{ave}. The form factor for a rectified sinusoid is $\pi/(2\sqrt{2}) \approx 1.11$. The *crest factor* is V_m/V_{eff}. The crest factor for a sinusoid is $\sqrt{2} \approx 1.41$.

ADMITTANCE (48-5)

$$\mathbf{Y} = \frac{1}{\mathbf{Z}}$$

RESISTORS (48-7)

governing equation: $V(t) = I(t)R$

power: $P(t) = \frac{1}{2} I_m V_m - \frac{1}{2} I_m V_m \cos 2\omega t$

average power: $P_{ave} = \frac{1}{2} I_m V_m = I_{eff} V_{eff}$

RESONANCE (48-10)

At resonance, the frequency is

$$\omega_0 = \frac{1}{\sqrt{LC}} = 2\pi f_0 \quad \text{[rad/s]}$$

Series Resonance (48-11)

$Z = R$
Z is minimum.
I is maximum.
P is maximum.

$$Q = \text{quality factor} = \frac{\omega_0 L}{R} = \frac{1}{\omega_0 C R} = \frac{1}{R}\sqrt{\frac{L}{C}}$$

$$\text{bandwidth} = \frac{\omega_0}{Q} = \frac{2\pi f_0}{Q}$$

Parallel Resonance (48-12)

$Z = R$
Z is maximum.
I is minimum.
P is minimum.

$$Q = \text{quality factor} = \omega_0 C R = \frac{R}{\omega_0 L} = R\sqrt{\frac{C}{L}}$$

$$\text{bandwidth} = \frac{\omega_0}{Q} = \frac{2\pi f_0}{Q}$$

COMPLEX POWER (48-14)

1 hp = 0.7457 kW
P = real power, kW
S = apparent power, kVA
Q = reactive power, kVAR
power factor = $\cos\phi$
reactive factor = $\sin\phi$

$$P = \tfrac{1}{2} I_{R,m} V_{R,m} = I_R V_R = S\cos\phi$$

$$Q = \tfrac{1}{2} I_{X,m} V_{X,m} = I_X V_X = S\sin\phi$$

$$S = \sqrt{P^2 + Q^2} = \tfrac{1}{2} I_{\text{line},m} V_{\text{line},m} = I_{\text{line,eff}} V_{\text{line,eff}}$$

POWER FACTOR CORRECTION (48-15)

The required change in reactive power to change the power angle from ϕ_1 to ϕ_2 is

$$\Delta Q = P(\tan\phi_1 - \tan\phi_2)$$

The capacitance (in F) required to change the reactive power by the above amount is given below. V_{line} is the maximum value of the sinusoid.

$$C = \frac{\Delta Q}{\pi f V_{\text{line},m}^2} = \frac{\Delta Q}{2\pi f V_{\text{line,eff}}^2}$$

SIMPLE TRANSFORMERS (48-17)

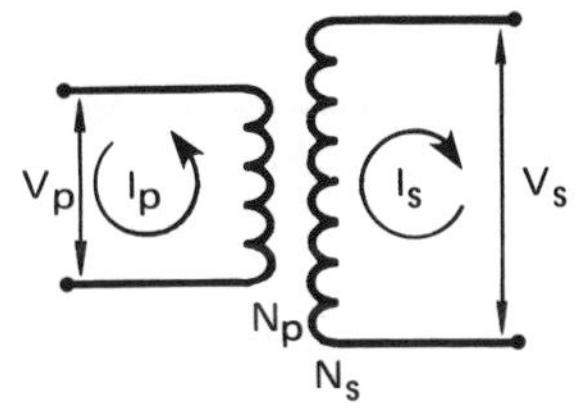

$$a = \frac{N_p}{N_s} = \frac{V_p}{V_s} = \frac{I_s}{I_p}$$

$a > 1$: step down

$a < 1$: step up

For impedance matching, the turns ratio should be

$$a^2 = \frac{Z_p}{Z_s}$$

mutual inductance (M):

$$M = k\sqrt{L_p L_s}$$

$k = 1$ for ideal transformers

3-PHASE LOADS (49-3)

Delta-Connected Loads

$$V_{\text{line}} = V_{\text{phase}}$$

$$I_{\text{line}} = \sqrt{3}\, I_{\text{phase}}$$

$$P_{\text{total}} = 3P_{\text{phase}} = 3V_{\text{phase}} I_{\text{phase}} \cos\phi = \sqrt{3}\, V_{\text{line}} I_{\text{line}} \cos\phi$$

Wye-Connected Loads

$$V_{\text{line}} = \sqrt{3}\, V_{\text{phase}}$$

$$I_{\text{line}} = I_{\text{phase}}$$

$$P_{\text{total}} = 3P_{\text{phase}} = 3V_{\text{phase}} I_{\text{phase}} \cos\phi = \sqrt{3}\, V_{\text{line}} I_{\text{line}} \cos\phi$$

GENERATED FREQUENCY (50-3)

$$f = \frac{(\text{rpm})(\text{number of poles})}{120}$$

SYNCHRONOUS MACHINES (50-4)

$$n_s = \text{synchronous speed} = \frac{120f}{\text{number of poles}} \quad [60\text{ Hz}]$$

INDUCTION MACHINES (50-4)

The stator field rotates at synchronous speed (n_s). If the rotor rotates at n_r, the *slip* is

$$s = \frac{n_s - n_r}{n_s} \times 100\%$$

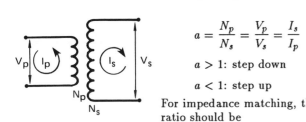

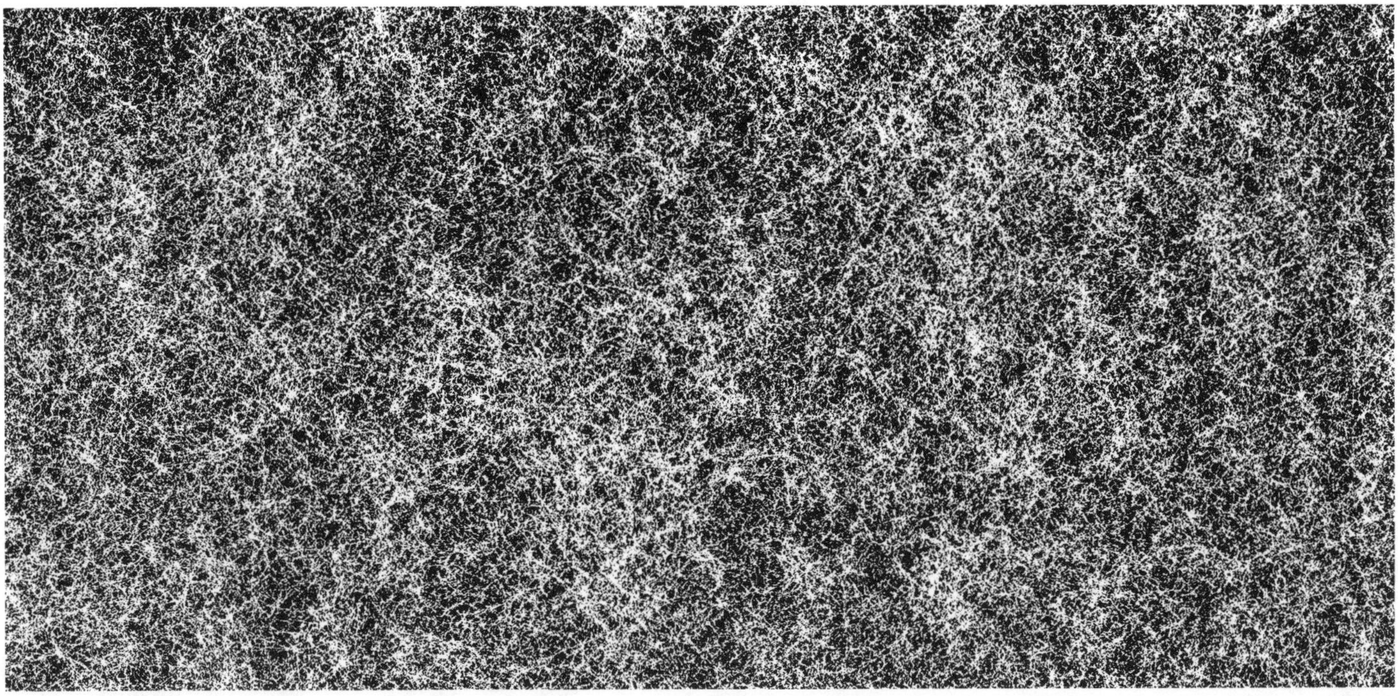

HEAT TRANSFER (Chap. 31)

Conduction (31-2)

Through a single layer: $q = \dfrac{kA\Delta T}{L}$

Through several layers: $q = \dfrac{A\Delta T}{\sum \frac{L_i}{k_i}}$

Convection (31-4)

Through a film: $q = hA\Delta T$

Through layers with films: $q = \dfrac{A\Delta T}{\sum \frac{L_i}{k_i} + \sum \frac{1}{h_j}}$

Logarithmic mean temperature difference:

$$\Delta T_m = \frac{\Delta T_A - \Delta T_B}{\ln \frac{\Delta T_A}{\Delta T_B}}$$

In a heat exchanger: $q = hA\Delta T_m$

Radiation (31-5)

The energy radiated from a hot body at temperature T (in degrees absolute) and with emissivity ϵ is

$$q = \sigma \epsilon A T^4$$

σ is the Stefan-Boltzmann constant, equal to 0.1713×10^{-8} BTU/hr-ft^2-°R^4 or 5.67×10^{-8} W/m^2·K^4.

$\epsilon = 1$ for blackbodies.

The net heat transfer between two bodies is

$$q = \sigma A_1 F_{1\text{-}2}(T_1^4 - T_2^4)$$

If object 1 is small and completely enclosed by object 2, then $F_{1\text{-}2} = \epsilon_1$.

LIGHT AND ILLUMINATION (Chap. 52)

Wavelength, λ, may be measured in microns (10^{-6} m), millimicrons (10^{-9} m), or angstroms, Å, (10^{-10} m).

The ***velocity of light***, c, in a vacuum is approximately 3×10^8 m/s, 9.84×10^8 ft/sec, and 6.71×10^8 mph. Velocity is constant although the frequency, f, will depend on the wavelength.

$$c = \lambda f$$

If the actual emitted frequency is f, and the relative separation velocity is v, the observed *Doppler* frequency is

$$f' = \frac{f\left(1 - \frac{\text{v}}{c}\right)}{\sqrt{1 - \left(\frac{\text{v}}{c}\right)^2}}$$

The ***illumination***, E, (in ft-candles) on a surface a distance r from an omnidirectional source radiating Φ lumens is

$$E = \frac{\Phi}{4\pi r^2}$$

OPTICS (52-8)

Total reflection from a transparent medium will occur when the incident angle exceeds the *critical angle*.

$$\phi = \arcsin\left(\frac{1}{n}\right)$$

n is known as the *index of refraction*.

$$n = \frac{c_{\text{vacuum}}}{c_{\text{medium}}} = \frac{\sin \phi_i}{\sin \phi_r}$$

The *lens equation*† relates the distance from the lens center to the object (o), image (i), and the focus (f).

$$\frac{1}{o} + \frac{1}{i} = \frac{1}{f}$$

The *magnification*† produced by a spherical lens is

$$M = -\frac{i}{o}$$

The *lens makers' equation* requires knowledge of the radii of curvature. Values of r are negative for concave surfaces.

$$\frac{1}{f} = (n-1)\left(\frac{1}{r_1} - \frac{1}{r_2}\right)$$

If f is measured in meters, $1/f$ is the *power of the lens* measured in units of diopters.

The *mirror equation*‡ relates the distance from the mirror to the object (o), image (i), focus (f), and center of curvature (r).

$$\frac{1}{o} + \frac{1}{i} = \frac{1}{f} = \frac{2}{r}$$

The *magnification*‡ produced by a spherical mirror is

$$M = -\frac{i}{o}$$

WAVES AND SOUND (Chap. 53)

Velocity, frequency, and wavelength are related.

$$\text{v} = f\lambda$$

Velocity of a longitudinal wave in a solid or liquid is

$$\text{v}_{\text{longitudinal}} = \sqrt{\frac{g_c E}{\rho}}$$

Velocity of a longitudinal wave in a gas is

$$\text{v}_{\text{longitudinal}} = \sqrt{k g_c R T}$$

Velocity of a transverse wave in a taut wire depends on the wire tension (F), length (L), and total mass (m_t).

$$\text{v}_{\text{transverse}} = \sqrt{\frac{F L g_c}{m_t}}$$

The frequency of the mth overtone (same as the $(m+1)$th harmonic) in a taut wire is

$$f = \frac{(m+1)\text{v}_{\text{transverse}}}{2L}$$

The *fundamental frequency* is the first harmonic, so $m = 0$.

---

† o is positive to the left of the lens, and i is positive to the right of the lens. f is negative for a diverging lens.

‡ f and r are negative for convex mirrors. Therefore, i will also be negative for convex mirrors.

RESPONSE VARIABLES (54-2)

A dependent variable that predicts the performance of a system is known as a *response variable.* Position, velocity, and acceleration vary with time and are the dependent response variables. Time is usually the independent variable.

FORCING FUNCTIONS (54-2)

An equation that describes the introduction of energy into the system as a function of time is known as a *forcing function.* The most common forcing functions used in the analysis of engineering systems are sinusoidal functions.

$$F(t) = \sin \omega t$$

NATURAL, FORCED, AND TOTAL RESPONSE (55-1)

Natural response is induced when energy is applied to an engineering system and subsequently is removed. The system is left alone and is allowed to do what it would do naturally, without the application of further disturbing forces. If a system is acted upon by a force which repeats at regular intervals, the system will move in accordance with that force. This is known as *forced response.* The natural and forced responses are present simultaneously in forced systems. The sum of the two responses is known as the *total response.*

TRANSFER FUNCTIONS (55-3)

The ratio of the system response (output) to the forcing function (input) is known as the *transfer function*, $T(t)$.

$$T(t) = \frac{r(t)}{f(t)}$$

Transfer functions generally are written in terms of the s variable. This is accomplished, if $T(t)$ is known, by taking the Laplace transform of the transfer function. The result is the *transform of the transfer function*, typically called the *transfer function.*

$$T(s) = \mathcal{L}(T(t)) = \mathcal{L}\left(\frac{r(t)}{f(t)}\right)$$

FEEDBACK (55-4)

The basic feedback system consists of a *dynamic unit*, a *feedback element*, a *pick-off point*, and a *summing point.* The summing point is assumed to perform positive addition unless a minus sign is present.

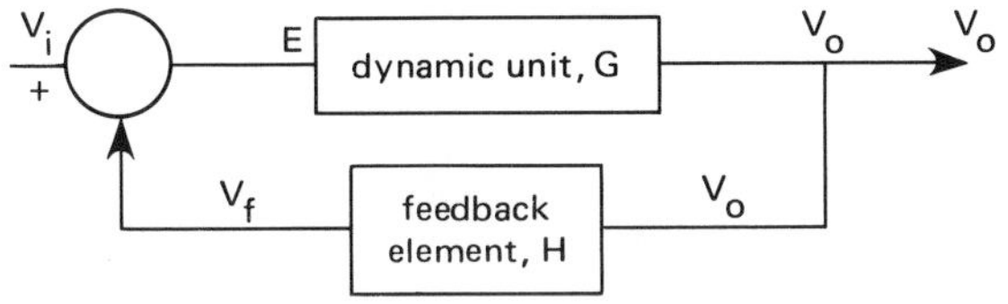

The dynamic unit transforms E into V_o according to the *forward transfer function*, G.

$$V_o = GE$$

For amplifiers, the forward transfer function, G, is known as the *direct* or *forward gain.* The sum of the input signal and the feedback is known as the *error.*

$$E = V_i + V_f$$

E/V_i is known as the *error ratio.* V_f/V_i is known as the *primary feedback ratio.*

The *pick-off point* transmits V_o back to the feedback element. The output of the dynamic unit is not reduced by the pick-off point. As the picked-off signal travels through the feedback loop, it is acted upon by the *feedback* or *reverse transfer function*, H.

The *closed-loop transfer function* (also known as the *control ratio* or the *system function*) is the ratio of the output to the signal.

$$G_{\text{loop}} = \frac{V_o}{V_i} = \frac{G}{1 - GH} \quad \text{[positive feedback]}$$

$$= \frac{G}{1 + GH} \quad \text{[negative feedback]}$$

The output of a feedback system is

$$V_o = GE = G_{\text{loop}} V_i$$

PREDICTING SYSTEM TIME RESPONSE (55-7)

Assuming that $T(s)$ and $\mathcal{L}(f(t))$ are known, the response function can be found by performing an inverse transformation.

$$r(t) = \mathcal{L}^{-1}(r(s)) = \mathcal{L}^{-1}(\mathcal{L}(f(t))T(s))$$

POLES AND ZEROS (55-8)

A *pole* of the transfer function is a value of s that makes $T(s)$ infinite. Specifically, a pole is a value of s that makes the denominator of $T(s)$ zero. A *zero* of the transfer function makes the numerator of $T(s)$ zero. Poles and zeros can be real or complex quantities. Poles and zeros can be repeated within a given transfer function—they need not be unique.

A rectangular coordinate system based on the real-imaginary axes is known as an *s-plane.* If poles and zeros are plotted on the *s-plane*, the result is a *pole-zero diagram.* Poles are represented on the pole-zero diagram as ×'s. Zeros are represented as ○'s.

PREDICTING SYSTEM TIME RESPONSE FROM POLE-ZERO DIAGRAMS (55-9)

Poles on the pole-zero diagram can be used to predict the usual response of engineering systems. Zeros are not used.

pure oscillation: Sinusoidal oscillation will occur if a pole-pair falls on the imaginary axis. A pole with a value of $\pm j\omega$ will produce oscillation with a natural frequency of ω rad/s.

exponential decay: Pure exponential decay is indicated when a pole falls on the real axis. A pole with a value of $-r$ will produce a decaying exponential e^{-rt} with time constant $1/r$.

damped oscillation: Decaying sinusoids result from pole-pairs in the second and third quadrants of the s-plane. A pole-pair having the value $r \pm j\omega$ will produce oscillation with natural frequency of

$$\omega_n = \sqrt{r^2 + \omega^2}$$

The closer the poles are to the real axis, the greater will be the damping effect. The closer the poles are to the imaginary axis, the greater will be the oscillatory effect.

STABILITY (55-11)

A pole with a value of $-r$ on the real axis corresponds to an exponential response of e^{-rt}. Similarly, a pole with a value of $+r$ on the real axis corresponds to an exponential response of e^{rt}. However, e^{rt} increases without limit. For that reason, such a pole represents instability.

Since any pole in the first and fourth quadrants of the s-plane will correspond to a positive exponential, a stable system must have poles limited to the left half of the s-plane (i.e., quadrants two and three).

ATOMIC AND NUCLEAR THEORY

MASS AND ENERGY EQUIVALENCE (61-2)

$$E = mc^2$$

E is in J.
m is in kg.
c is the velocity of light (3×10^8 m/s).

NUCLEAR CONVERSIONS (61-2)

Mass and Energy Conversion Factors*

(multiply ↓ to obtain →)

	kg	amu	J	eV
kg	1	6.0225×10^{26}	8.9876×10^{16}	5.6099×10^{35}
amu	1.6604×10^{-27}	1	1.4923×10^{-10}	9.3148×10^{8}
J	1.1127×10^{-17}	6.693×10^{9}	1	6.2418×10^{18}
eV	1.7826×10^{-36}	1.0736×10^{-9}	1.6021×10^{-19}	1

*based on $c = 2.997925 \times 10^8$ m/s
1 eV $= 1.60210 \times 10^{-19}$ J
1 amu $= 1.66043 \times 10^{-27}$ kg
1 kg $= 8.98755 \times 10^{16}$ J

BASIC ATOMIC PARTICLES (61-4)

(based on C-12 = 12)

particle	rest mass (kg)	rest mass (u)	charge†
alpha, α	$6{:}6416 \times 10^{-27}$	4.00153	+2
beta, β, e	9.1091×10^{-31}	0.000548597	−1
deuteron, d	3.3421×10^{-27}	2.01355	+1
electron, e	9.1091×10^{-31}	0.000548597	−1
gamma ray, γ	0	0	0
neutrino, ν	0	0	0
neutron, n	1.6748×10^{-27}	1.0086654	0
positron, +e	9.1091×10^{-31}	0.000548597	+1
proton, p	1.6725×10^{-27}	1.00727663	+1

†Multiply charge given by 1.60210×10^{-19} to obtain coulombs.
Multiply mass units (u) by 1.660×10^{-27} to obtain kilograms.

RADIOACTIVE DECAY (61-6)

$$m_t = m_o e^{-\lambda t}$$

$$A_t = \text{activity at time } t = A_o e^{-\lambda t} = \lambda N_t$$

$$\lambda = \text{decay constant} \approx \frac{0.6931}{t_{\frac{1}{2}}}$$

$$t_{\frac{1}{2}} = \text{half-life} \approx \frac{0.6931}{\lambda}$$

$$\begin{array}{c}\text{mean atom}\\ \text{life expectancy}\end{array} = \frac{1}{\lambda}$$

$$1 \text{ curie} = 3.7 \times 10^{10} \text{ disintegrations/sec}$$

SERIES OF HYDROGEN SPECTRA (61-8)

series name	series type
Lyman	ultraviolet
Balmer	visible
Paschen	infrared
Brackett	infrared
Pfund	infrared

BOHR'S ATOM (61-8)

first postulate: Electrons circle the nucleus in stable, nonradiating orbits.

second postulate: Energy is quantized.

The second postulate can be written as

$$E_2 - E_1 = h\nu$$

h is Planck's constant which has a value of 6.626×10^{-34} J·s, or 4.136×10^{-15} eV·s.

third postulate: Classical mechanics does not hold for electrons between orbits.

fourth postulate: Angular momentum is quantized.

The fourth postulate can be written as

$$mvr = \frac{nh}{2\pi}$$

n is the *principle quantum number*, with a value of 1 for the lowest electron orbit.

WAVE NATURE OF AN ELECTRON (61-9)

The de Broglie wave velocity is

$$\mathrm{v}_w = \nu f = \frac{h\nu}{m\mathrm{v}} = \frac{h\nu}{p}$$

The de Broglie wavelength depends on the mean electron radius and the principle quantum number.

$$\lambda = \frac{2\pi r}{n}$$

HEISENBERG'S UNCERTAINTY PRINCIPLE (61-9)

$$\epsilon_x \epsilon_p \sim h$$

$$\epsilon_t \epsilon_{\Delta E} \sim h$$

ϵ is the error or uncertainty in a quantity.
x is position in m.
p is momentum (mv) in kg·m/s.
t is time in s.
E is energy in J.
h is Planck's constant (6.626×10^{-34} J·s).

NUMBER OF ELECTRONS IN A SHELL (61-10)

n	shell name	capacity
1	K	2
2	L	8
3	M	18
4	N	32
5	O	50
6	P	72
7	Q	98

NUMBER OF ELECTRONS IN AN ORBITAL (61-10)

All orbitals have a maximum capacity of 2 electrons.

NUMBER OF ELECTRONS IN A SUBSHELL (61-11)

l	subshell name	capacity
0	s	2
1	p	6
2	d	10
3	f	14
4	g	18
5	h	22
6	i	26

PAULI EXCLUSION PRINCIPLE (61-11)

No two electrons may have the same set of quantum numbers. Stated another way, no two electrons may occupy the same orbital and both have the same spin.

RELATIVITY (61-16 to 61-18)

$$k = \frac{1}{\sqrt{1 - \left(\frac{v}{c}\right)^2}}$$

$$m_v = km_0 \quad \text{[mass increase]}$$

$$L_v = \frac{L_0}{k} \quad \text{[length contraction]}$$

$$t_v = kt_0 \quad \text{[time dilation]}$$

$$\nu_v = \nu_0 k \left(1 - \frac{v}{c}\right) \quad \text{[frequency]}$$

LINEAR ACCELERATORS (61-17)

The total energy possessed by a particle traveling at velocity v is the sum of its rest mass energy and its kinetic energy.

$$k = \frac{1}{\sqrt{1 - \left(\frac{v}{c}\right)^2}}$$

$$E_{\text{rest mass}} = m_0 c^2$$

$$E_{\text{kinetic}} = m_0(k-1)c^2$$

$$E_{\text{total}} = E_{\text{rest mass}} + E_{\text{kinetic}} = m_0 k c^2 = m_v c^2$$

$$m_v = km_0$$

$$v = c\sqrt{1 - \left(\frac{m_0}{m_v}\right)^2} = c\sqrt{1 - \left(\frac{1}{k}\right)^2}$$

CYCLOTRONS (61-18)

$$r = \text{path radius} = \frac{mv}{Bq}$$

$$\omega = \text{angular velocity} = \frac{v}{r} = \frac{Bq}{m}$$

$$v = \text{linear velocity} = \frac{qBr}{m}$$

$$T = \text{time per revolution} = \frac{2\pi r}{v} = \frac{2\pi m}{qB} = \frac{2\pi}{\omega}$$

$$E = \text{kinetic energy} = \tfrac{1}{2}mv^2 = \tfrac{1}{2}m(r\omega)^2 = \frac{(qBr)^2}{2m}$$

$$f = \text{required switching frequency} = \frac{1}{T} = \frac{qB}{2\pi m}$$

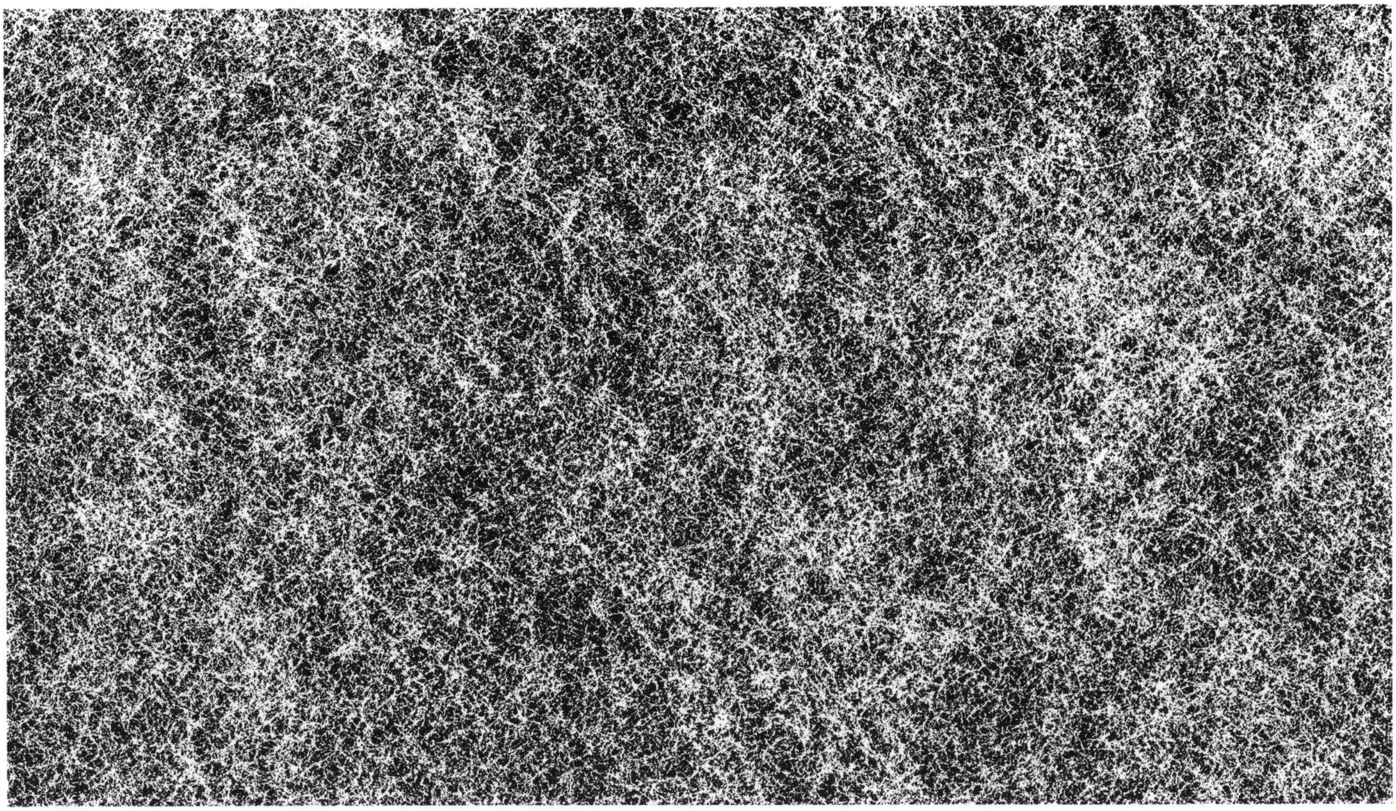

INDEX

A

acceleration, 28
  formulas, uniform, 28
  gravitational, 3
  uniform, 28
accelerators, linear, 37
acid/base neutralization, 18
ACRS, 9
adiabatic process, 14
admittance, 32
air
  atmospheric, definitions, 15
  composition of, 19
  density of, 1
  dry atmospheric, composition of, 15
  properties of, 12, 14
  ratio of specific heats for, 1
  specific gas constant for, 1
allowable stress, 26
alternatives, comparing, 8
aluminum
  designations, 22
  properties of, 25
  tempers, 22
Amagat law, 14
amortization, 8
amount
  annual, 8
  compound, 8
annual
  amount, 8
  effective interest rate, 8
annuity, 8
approximations, series, 6
arc length of a circular sector, 6
Archimedes' principle, 10
area
  between two curves, 7
  cross-sectional, 6
  moment of inertia, 1, 21
  of a circle, 1, 6
  of a circular sector, 6
  of a triangle, 1, 5
  of an ellipse, 6
  of right circular cone, 6
  of right circular cylinder, 6
  of sphere, 6
areas, surface, 6
arithmetic
  mean, 7
  series, 4
atmospheric
  air definitions, 15
  air, dry, composition of, 15
  pressure, 1, 10
atom, Bohr's, 36
atomic particles, basic, 36
atoms in a cell, number of, 22
average values, 32
Avogadro's
  law, 14
  number, 1

B

banking, roadway, 29
basic atomic particles, 36
basic identities, 5
beam
  deflections, simple, 26
  slope/deflection relations, 26
beams
  bending stress in, 26
  shear stress in, 26
bending stress in beams, 26
benefit/cost ratio, 9
Bernoulli equation, 11
binomial
  probability distribution, 7
  series, 4
Bohr's atom, 36
boiling point/freezing point changes, 18
book value, 8, 9
bridge, Wheatstone, 31
bulk modulus, 10
buoyancy, 10

C

cables
  catenary, 20
  parabolic, 20
capacitors, 32
capital recovery, 8
capitalized cost, 9
Carnot cycle
  COP, 17
  efficiency, 16
cash flow diagrams, 8
catenary cables, 20
cell
  number of atoms, 22
  packing parameters, 22
centrifugal force, 29
centroids, 21
  of composite (built-up) shapes, 21
changes, boiling point/freezing point, 18
channel, open, flow, 12
charge, 3, 30
Chezy equation, 12
circle, 1
  area moment of inertia for, 1
  area of, 1, 6
  circumference of, 1, 6
  polar moment of inertia for, 1
  resistive, power, 30
circular sector
  arc length of, 6
  area of, 6
circumference of a circle, 1, 6
closed system, 14
closed-loop transfer function, 35
coefficient
  of performance (COP), 17
  of restitution, 29
collisions, 29
columns in compression, slender, 26
combinations, permutations and, 7
combined stress, 25
combustion, heat of, 17
common metals, 22
comparing alternatives, 8
complex numbers, 4
complex power, 33
components of forces, 20
composite (built-up) shapes, centroids of, 21
composition of air, 15, 19
compound amount, 8
compounding, nonannual, 9
compressed gas equation of state, 14
compression, slender columns in, 26
concentration, units of, 18
conditions for equilibrium, 20
conduction, 34
cone, right circular, 6
conic sections, 6
conservation of energy, 11
constant
  equilibrium, 18
  fundamental, 1, 3
  gravitational, 1
  ionization, 18
  Joule's, 1
  Stefan-Boltzmann, 34
  universal gas, 1, 13
constant pressure processes (closed system, ideal gas), 14
constant temperature processes (closed system, ideal gas), 15
constant value dollars, 9
constant volume processes (closed system, ideal gas), 14
consumer loans, 9
continuity equation, 11
control ratio, 35
convection, 34
convention, year-end, 8
conversion, energy, work, and power, 17
conversions, 13, 28
  energy, 30
  nuclear, 36
  power, 30
  sine-cosine, 32
  temperature, 1
coordinate systems, 5
COP, 17
correction, power factor, 33
cosines, law of, 5
cost, 8
  capitalized, 9
  equivalent uniform annual (EUAC), 8, 9
  sunk, 8
Coulomb's law, 30
couples, 20
creep
  rate, 24
  strength, 24
  test, 24
crest factor, 32
cross product, 5
cross-sectional areas, 6
crystalline directions, 23
crystalline planes, 23
  spacing between, 23
crystalline structure, 22
current, house, frequency of, 1
curves, area between two, 7
cycle
  Carnot, COP, 17
  Carnot, efficiency, 16
  power, 16
  Rankine, with superheat, 16
  refrigeration, 17
  thermal efficiency of, 16
cyclotrons, 37
cylinder
  mass moment of inertia of, 1
  right circular, 6

D

Dalton's law, 14
damped oscillation, 35
Darcy friction factor, 11
data, derived, 1
DC transient solutions, 31
decay
  exponential, 35
  radioactive, 36
decibels, 1, 30

INDEX

declining balance depreciation rate, 8
defects, point, 23
definitions, atmospheric air, 15
deflections, simple beam, 26
delta-connected loads, 33
delta-wye transformations, 31
density, 3, 10
  flux, 30
  of air, 1
  of mercury, 1
  of water, 1
depreciation, 8, 9
  double declining balance, 9
  rate, declining balance, 8
  sinking fund, 9
  straight line, 9
  sum of the years' digits, 9
derivative operations, 6
derived data, 1
designations
  aluminum, 22
  steel, 22
determinants, 4
determinate trusses, 20
deviation
  sample standard, 7
  standard, 7
dewpoint temperature, 15
diagram
  cash flow, 8
  moment, 26
  pole-zero, 35
  shear, 25
diameter
  of a pipe flowing full, 12
  of a pipe flowing half-full, 12
differentiation, extrema by, 6
diffraction, x-ray, 23
diffusion, Fick's laws of, 23
direct gain, 35
direct reduction loans, 9
directions, crystalline, 23
discounting factors, 8
distance, 3, 28
distribution
  binomial probability, 7
  Poisson probability, 7
dollars, constant value, 9
Doppler frequency, 34
dot product, 5
double declining balance, 9
drag, 12
dry-bulb temperature, 15
ductility, 23
dynamic unit, 35

E

eccentric normal stress, 26
economic indicator, 9
effective interest rate per period, 8
effective values, 32
efficiency, 16
elasticity, modulus of, 1, 23
electric field, 30
electrolysis, Faraday's law of, 19
electrons
  in a shell, number of, 36
  in a subshell, number of, 37
  in an orbital, number of, 37
  wave nature of, 36
electrostatics, 30
element
  and radical table, 18
  feedback, 35
ellipse, area of, 6
elongation
  at fracture, percent, 24
  under normal stress, 25
end restraint coefficients, theoretical, 26
endurance limit, 24
endurance stress, 24
energy, 28
  and mass equivalence, 36
  conservation of, 11
  conversions, 17, 30
  kinetic, 28
  potential, 28
  system potential, 30
engineering strain, 23
engineering stress, 23
enthalpy of reaction, 18
equal series present worth, 8
equation
  Bernoulli, 11
  Chezy, 12
  continuity, 11
  Euler's, 4
  Hazen-Williams, 12
  lens, 34
  lens makers', 34
  Manning, 12
  mirror, 34
  of a straight line, 5
  of state, compressed gas, 14
  of state for ideal gases, 13
  quadratic, 4
  Torricelli, 11
equilibrium
  conditions for, 20
  constants, 18
equivalence, mass and energy, 36
equivalent
  Norton, 31
  Thevenin, 31
  uniform annual cost (EUAC), 8, 9
  weight, 18
error, 35
  ratio, 35
EUAC, 8
Euler's equation, 4
exclusion principle, Pauli, 37
exponential decay, 35
exponents, 4
extrema by differentiation, 6

F

factor
  crest, 32
  discounting, 8
  form, 32
Faraday's law, 30
  of electrolysis, 19
fatigue
  life, 24
  limit, 24
  test, 24
feedback, 35
  element, 35
  ratio, primary, 35
  transfer function, 35
Fick's laws of diffusion, 23
field
  electric, 30
  magnetic, 30
first law
  mathematical formulation of, 13
  of thermodynamics, 13
  sign convention, 13
flow
  open channel, 12
  through nozzles, 16
fluid, speed of sound in, 10
flux, 30
  density, 30
force, centrifugal, 29
forced response, 35
forces, 20
  components of, 20
  in vector form, 20
  truss, sign conventions for, 20
forcing functions, 35
form factor, 32
formula
  PLAN (internal combustion engines), 16
  Taylor's, 6
  two-angle, 5
  uniform acceleration, 28
forward gain, 35
forward transfer function, 35
fraction weighted gas mixtures, 14
frequency, 32
  Doppler, 34
  fundamental, 34
  generated, 33
  of house current, 1
friction, 21
  factor, Darcy, 11
  loss, 11
Froude number, 12
fuel consumption
  rate of, 16
  specific, 16
function, system, 35
functions
  forcing, 35
  hyperbolic, 5
  transfer, 35
fund, uniform series sinking, 8
fundamental constants, 1, 3
fundamental frequency, 34
fusion, latent heat of, 13
future worth, 8

G

gain, 35
galvanic series, 19
gas
  compressed, equation of state, 14
  constant for air, specific, 1
  constant, universal, 1, 13
  ideal, equations of state for, 13
  ideal, mixtures of, 14
  ideal, process law, 14
  ideal, specific heat relationships for, 14
  laws, 14
  mixtures, 14
  molecular weight of, 18
general form, 5
general plane surface, pressure on, 10
generated frequency, 33
geometric
  mean, 7
  series, 4
gradient amount, uniform, 8
gradient, uniform, 8
gravimetrically weighted gas mixtures, 14

INDEX

gravitational acceleration, 3
gravitational constant, 1
gravity, 1
gyration, radius of, 21, 28

H

hardness versus ultimate strength, 24
harmonic mean, 7
harmonic motion, simple, 29
Hazen-Williams equation, 12
head, 1
heat
  latent, 13
  of combustion, 17
  of reaction, 18
  pumps, 17
  sensible, 13
  specific, 1
  transfer, 34
heating value, 17
Heisenberg's uncertainty principle, 36
higher heating value, 17
Hooke's law, 25, 26
horsepower
  hydraulic, 12
  required to maintain velocity, 28
  versus torque, 16
humidity, 15
hydraulic horsepower, 12
hydrogen spectra, series of, 36
hydrostatic pressure on a flat plane, 10
hyperbolic
  functions, 5
  identities, 5

I

ice, properties of, 13
ideal gas process law, 14
ideal gases
  equations of state for, 13
  mixtures of, 14
  specific heat relationships for, 14
identities
  basic, 5
  hyperbolic, 5
  logarithm, 4
illumination, 34
impact test, 24
impulse-momentum, 11
  principle, 29
inclined plane surface, pressure on, 10
income tax, 8, 9
index
  of refraction, 34
  price, 9
indicator, economic, 9
indices, Miller, 23
inductance, mutual, 33
induction machines, 33
inductors, 32
inertia
  area moment of, 1, 21
  mass moment of, 1
  polar moments of, 1, 21
inflation, 9
inflection point, 6
instantaneous relationship between rotational and linear motion, 29
integral operations, 7
intercept form, 5
interest
  rate, 8
  simple, 9
investment, return on (ROI), 8
ionization constants, 18
isentropic
  efficiency of a pump, 16
  efficiency of a turbine, 16
  process, 14, 15
isobaric process, 14
isochoric process, 14
isometric process, 14
isothermal process, 14

J

joints, 20
Joule's constant, 1

K

kinetic energy, 28
Kirchhoff's laws, 31

L

latent heat, 13
law
  Amagat, 14
  Avogadro's, 14
  Coulomb's, 30
  Dalton's, 14
  Faraday's, 30
  gas, 14
  Hooke's, 25, 26
  Kirchhoff's, 31
  Newton's, 29
  of cosines, 5
  of sines, 5
  Ohm's, 30
  pump affinity/similarity, 12
lens
  equation, 34
  makers' equation, 34
  power of, 34
life, fatigue, 24
lift, 12
light, 34
  speed of, 1
  velocity of, 34
limit
  endurance, 24
  fatigue, 24
linear accelerators, 37
lines, perpendicular, slopes of, 6
liquid-vapor mixture, 13
loads, 33
loans, 9
logarithm
  identities, 4
  mean temperature difference, 34
loss, friction, 11
losses, minor, 11

M

Mach number, 10
machines
  induction, 33
  synchronous, 33
MACRS, 9
magnetic field, 30
magnetism, 30
magnification, 34
Manning equation, 12
manometers, 10
MARR, 8
mass, 3
  and energy equivalence, 36
  moment of inertia, 1, 28
  weighted gas mixtures, 14
mathematical formulation of the first law, 13
mean
  arithmetic, 7
  geometric, 7
  harmonic, 7
mensuration, 6
mercury
  density of, 1
  properties of, 12
metals, common, 22
meter, venturi, 11
Miller indices, 23
minimum attractive rate of return (MARR), 8
minor losses, 11
mirror equation, 34
mixtures of ideal gases, 14
modulus
  bulk, 10
  of elasticity, 23
  of elasticity for steel, 1
  of shear for steel, 1
  of toughness, 23
  shear, 23
molality, 18
molarity, 18
molecular weight, 1
  of gases, 18
moles, 18
moment diagrams, 26
moment of inertia
  area, 1, 21
  mass, 1
  polar, 1, 21
moments, 20
motion, instantaneous relationship between rotational and linear, 29
mutual inductance, 33

N

natural response, 35
neutralization, acid/base, 18
Newton's laws, 29
nominal annual interest rate, 8
nonannual compounding, 9
normal stress, 25
normality, 18
Norton equivalent, 31
nozzles, flow through, 16
nuclear conversions, 36
number
  Avogadro's, 1
  Froude, 12
  Mach, 10
  Reynolds, 11
number of atoms in a cell, 22
number of electrons
  in a shell, 36
  in a subshell, 37
  in an orbital, 37
numbers, complex, 4

O

Ohm's law, 30
open channel flow, 12
open system, 14

operations
derivative, 6
integral, 7
optics, 34
orbital, number of electrons, 37
orifice plate, 11
oscillation
damped, 35
pure, 35
oxidation-reduction reactions, 18

P

parabolic cables, 20
parallel axis theorem, 21, 28
parallel, resistors, 30
parameters, cell packing, 22
parts per million (ppm), 18
Pauli exclusion principle, 37
percent elongation at fracture, 24
performance, coefficient of (COP), 17
period, 32
permutations and combinations, 7
perpendicular lines, slopes of, 6
pH, 18
physical properties, 1
pick-off point, 35
pipe
diameter of, 12
stress, 27
PLAN formula (internal combustion engines), 16
planes, crystalline, 23
plate, orifice, 11
pOH, 18
point
defects, 23
inflection, 6
pick-off, 35
-slope form, 5
summing, 35
Poisson probability distribution, 7
Poisson's ratio, 25
polar moments of inertia, 21
for a circle, 1
pole-zero diagram, 35
poles, 35
polynomials, 4
polytropic process, 14
potential, 30
potential energy, 28
power, 28
complex, 33
conversions, 17, 30
cycle, 16
factor correction, 33
in a resistive circuit, 30
of the lens, 34
ppm, 18
prefixes, 30
present worth, 8
equal series, 8
pressure, 1, 3
atmospheric, 1, 10
hydrostatic, on a flat plane, 10
on general plane surface, 10
on inclined plane surface, 10
on submerged plane surface, 10
price index, 9
primary feedback ratio, 35
principal stresses, 25
principle
Archimedes', 10
Heisenberg's uncertainty, 36
impulse-momentum, 29
Pauli exclusion, 37
work-energy, 28
principle quantum number, 36
processes, 14, 15
product
cross, 5
dot, 5
projectile motion, 29
properties
of a liquid-vapor mixture, 13
of air, 14
of aluminum, 25
of ice, 13
of mercury, 12
of standard air, 12
of steel, 25
of water, 13
of water at room temperature, 12
physical, 1
pump
affinity/similarity laws, 12
isentropic efficiency, 16
pumps, heat, 17
pure oscillation, 35

Q

quadratic equation, 4
quality of a liquid-vapor mixture, 13
quantum number, principle, 36

R

radiation, 34
radical and element table, 18
radioactive decay, 36
radius of gyration, 21, 28
Rankine cycle with superheat, 16
rate
creep, 24
of fuel consumption, 16
of return (ROR), 8
of return, minimum attractive rate (MARR), 8
ratio
benefit/cost, 9
control, 35
error, 35
humidity, 15
of specific heats for air, 1
Poisson's, 25
primary feedback, 35
reaction, enthalpy (heat), 18
reactions, oxidation-reduction, 18
recovery, capital, 8
rectangle, area moment of inertia for, 1
refraction, index of, 34
refrigeration, 17
relations, beam slope/deflection, 26
relationship between shear and elastic moduli, 25
relative humidity, 15
relativity, 37
resistive circuit, power, 30
resistivity, 30
resistors, 32
in series and parallel, 30
resonance, 32
response
natural, forced, and total, 35
variables, 35
restitution, coefficient of, 29
resultants, 20
return
minimum attractive rate of (MARR), 8
on investment (ROI), 8
rate of (ROR), 8
reverse transfer function, 35
reversible adiabatic processes (closed system, ideal gas), 15
Reynolds number, 11
right circular cone, 6
right circular cylinder, 6
roadway banking, 29
ROI, 8
root-mean-square value, 7, 32
ROR, 8
rotational
motion, 29
speed, 1
systems, mass moment of inertia, 28
systems, Newton's second law for, 29
rupture strength, 24

S

s-plane, 35
salvage value, 8
sample
standard deviation, 7
variance, 7
scales, temperature, 13
second law of thermodynamics, 13
sections, conic, 6
sensible heat, 13
series
approximations, 6
arithmetic, 4
binomial, 4
galvanic, 19
geometric, 4
of hydrogen spectra, 36
resistors, 30
shear
and elastic moduli, relationship between, 25
diagrams, 25
modulus, 23
modulus of, for steel
stress, 25
stress in beams, 26
shell, number of electrons in, 36
sign convention, first law, 13
sign conventions for truss forces, 20
similarity, 12
simple beam deflections, 26
simple harmonic motion, 29
simple interest, 9
simple transformers, 33
sine-cosine conversions, 32
sines, law of, 5
single payment compound amount, 8
sinking fund, 9
slender columns in compression, 26
slip, 33
slope-intercept form, 5
slopes of perpendicular lines, 6
sound, 34
speed, in a fluid, 10
spacing between crystalline planes, 23
specific fuel consumption, 16
specific gas constant for air, 1

INDEX

specific gravity, 1, 10
specific heat relationships for ideal gases, 14
specific heats, 1
  ratio for air, 1
specific humidity, 15
speed
  of light, 1
  of sound in a fluid, 10
  rotational, 1
sphere
  area of, 6
  volume of, 1, 6
springs, 26
  systems of, 28
stability, 35
standard deviation, 7
standard gravity, 1
standard temperature and pressure (STP), 13
static-pitot tube, 11
statistics, 7
steel
  designations, 22
  modulus of elasticity for, 1
  modulus of shear for, 1
  properties of, 25
Stefan-Boltzmann constant, 34
stiffness, 28
STP, 13
straight line
  depreciation, 9
  equations of, 5
strain, 25
  engineering, 23
  thermal, 25
  torsional, in a shaft, 27
  true, 23
strength
  creep, 24
  rupture, 24
stress, 25
  allowable, 26
  bending, in beams, 26
  combined, 25
  eccentric normal, 26
  endurance, 24
  engineering, 23
  normal, 25
  pipe, 27
  shear, 25
  shear, in beams, 26
  thermal, 25
  torsional, in a shaft, 27
  true, 23
stresses, principal, 25
structure, crystalline, 22
sublimation, latent heat of, 13
submerged plane surface, pressure on, 10
subshell, number of electrons, 37
sum of the years' digits, 9
summing point, 35
sunk costs, 8
surface areas, 6
synchronous machines, 33
system
  closed, 14
  coordinate, 5
  function, 35
  of springs, 28
  open, 14
  potential energy, 30
  time response, 35

T

table, element and radical, 18
tanks, thin-walled, 27
tax, income, 8, 9
Taylor's formula, 6
temperature, 3
  and pressure, standard, 13
  conversions, 1
  dewpoint, 15
  difference, logarithmic mean, 34
  dry-bulb, 15
  scales, 13
  transition, 24
  wet-bulb, 15
tempers, aluminum, 22
tensile test, 23
test
  creep, 24
  fatigue, 24
  impact, 24
  tensile, 23
  toughness, 24
theorem, parallel axis, 21, 28
theoretical end restraint coefficients, 26
thermal efficiency of the entire cycle, 16
thermal stress and strain, 25
thermodynamic relationships for any process (ideal gases), 14
thermodynamic systems, types of, 14
thermodynamics, laws of, 13
Thevenin equivalent, 31
thin-walled tanks, 27
third law of thermodynamics, 13
three-phase loads, 33
throttling process, 14
time, 32
time response, system, 35
titration, 18
torque, horsepower versus, 16
Torricelli equation, 11
torsional stress and strain in a shaft, 27
total response, 35
toughness
  modulus of, 23
  test, 24
transfer
  function, 35
  heat, 34
transform of the transfer function, 35
transformations, delta-wye, 31
transformers, simple, 33
transient solutions, DC, 31
transition temperature, 24
triangle, area of, 1, 5
trigonometry, 5
triple point of water, 1
true strain, 23
true stress, 23
truss forces, sign conventions for, 20
trusses, determinate, 20
tube, static-pitot, 11
turbine, isentropic efficiency, 16
two-angle formulas, 5
types of processes, 14
types of thermodynamic systems, 14

U

ultimate strength, hardness versus, 24
uncertainty principle, Heisenberg's, 36
uniform acceleration, 28
  formulas, 28
uniform electric fields, 30
uniform gradient, 8
  amount, 8
uniform series sinking fund, 8
unit, dynamic, 35
units
  of concentration, 18
  refrigeration, 17
universal gas constant, 1, 13

V

value
  book, 8, 9
  heating, 17
  higher heating, 17
  root-mean-square, 7, 32
  salvage, 8
values, 32
vaporization, latent heat of, 13
variables, response, 35
variance, 7
  sample, 7
vector form, forces in, 20
vectors, 4
velocity, 3, 28
  horsepower required to maintain, 28
  of light, 34
venturi meter, 11
viscosity, 10
volume, 3, 6
  of a sphere, 1, 6
  of right circular cone, 6
  of right circular cylinder, 6
volumetrically weighted gas mixtures, 14

W

water
  at room temperature, properties of, 12
  density of, 1
  properties of, 13
  triple point of, 1
wave nature of an electron, 36
waves, 34
weight
  equivalent, 18
  molecular, 1
  of gases, molecular, 18
wet-bulb temperature, 15
Wheatstone bridge, 31
work, 28, 30
  conversions, 17
work-energy principle, 28
worth
  equal series present, 8
  future, 8
  present, 8
wye-connected loads, 33

X

x-ray diffraction, 23

Y

year-end convention, 8

Z

zeros, 35
zeroth law of thermodynamics, 13

Quick – I need additional study materials!

Please rush me the review materials I have checked. I understand any item may be returned for a full refund within 30 days. I have provided my bank card number as method of payment, and I authorize you to charge your current prices against my account.

	Solutions Manuals:
For the E-I-T Exam:	
[ ] Engineer-In-Training Reference Manual	[ ]
[ ] Engineering Fundamentals Quick Reference Cards	
[ ] Engineer-In-Training Sample Examinations	
[ ] E-I-T Mini-Exams	
[ ] 1001 Solved Engineering Fundamentals Problems	
[ ] E-I-T Review: A Study Guide	
For the P.E. Exams:	
[ ] Civil Engineering Reference Manual	[ ]
[ ] Civil Engineering Sample Examination	
[ ] Civil Engineering Quick Reference Cards	
[ ] Seismic Design of Building Structures	
[ ] Timber Design for the Civil P.E. Examination	
[ ] Mechanical Engineering Reference Manual	[ ]
[ ] Mechanical Engineering Quick Reference Cards	
[ ] Mechanical Engineering Sample Examination	
[ ] Electrical Engineering Reference Manual	[ ]
[ ] Chemical Engineering Reference Manual	[ ]
Recommended for all Exams:	
[ ] Expanded Interest Tables	
[ ] Engineering Unit Conversions	

SHIP TO:

Name ______________________

Company ______________________

Street ______________________ Apt. No. ________

City ______________ State _____ Zip ________

Daytime phone number ______________________

CHARGE TO (required for immediate processing):

______________________ ________

VISA/MC/AMEX account number — expiration date

____

name on card

____

signature

Send more information.

Please send me descriptions and prices of all available E-I-T and P.E. review books. I understand there will be no obligation on my part.

Name ______________________

Address ______________________

City ______________________

State ______________ Zip ________

A friend of mine is taking the exam, too. Send additional literature to:

Name ______________________

Address ______________________

City ______________________

State ______________ Zip ________

I have a comment . . .

□ I think you should add the following subject to page ____________ .

□ I think there is an error on page ____________ . Here is the way I think it should be:

Title of this book: **ENGINEERING FUNDAMENTALS QUICK REFERENCE CARDS, Fourth Edition**

Contributed by (optional): [ ] Please tell me if I am correct.

Name ______________________

Address ______________________

City ______________________

NO POSTAGE NECESSARY IF MAILED IN THE UNITED STATES

BUSINESS REPLY MAIL

FIRST CLASS PERMIT NO. 33, BELMONT, CA

POSTAGE WILL BE PAID BY ADDRESSEE

PROFESSIONAL PUBLICATIONS, INC.
1250 Fifth Avenue
Belmont, CA 94002-9900

NO POSTAGE NECESSARY IF MAILED IN THE UNITED STATES

BUSINESS REPLY MAIL

FIRST CLASS PERMIT NO. 33, BELMONT, CA

POSTAGE WILL BE PAID BY ADDRESSEE

PROFESSIONAL PUBLICATIONS, INC.
1250 Fifth Avenue
Belmont, CA 94002-9900

NO POSTAGE NECESSARY IF MAILED IN THE UNITED STATES

BUSINESS REPLY MAIL

FIRST CLASS PERMIT NO. 33, BELMONT, CA

POSTAGE WILL BE PAID BY ADDRESSEE

PROFESSIONAL PUBLICATIONS, INC.
1250 Fifth Avenue
Belmont, CA 94002-9900